MACHINE SHOP PROJECTS

Price $2.00

The projects illustrated and described in this book are intended to help students in vocational and industrial schools in learning basic machining skills. Project 13, 1″ Bolt and Nut, is a typical example. The drawing shows the assembly and details of the bolt and nut and the description outlines the operations in the sequence that the student follows. The procedure is similar to that used in industrial plants throughout the country, not only in training apprentices but in regular production work.

The book "How to Run a Lathe" is to be used in conjunction with this machine shop project book. It gives necessary detailed instructions for the various operations called for in the project book.

If the instructions are followed carefully it will be but a short time before the student who shows an aptitude for the work is qualified to enter the shop as a learner or helper to a machinist, toolmaker or die maker.

CONTENTS

Reprinted by Lindsay Publications Inc
Bradley IL 60915

ISBN 1-55918-248-2

5 6 7 8 9 0

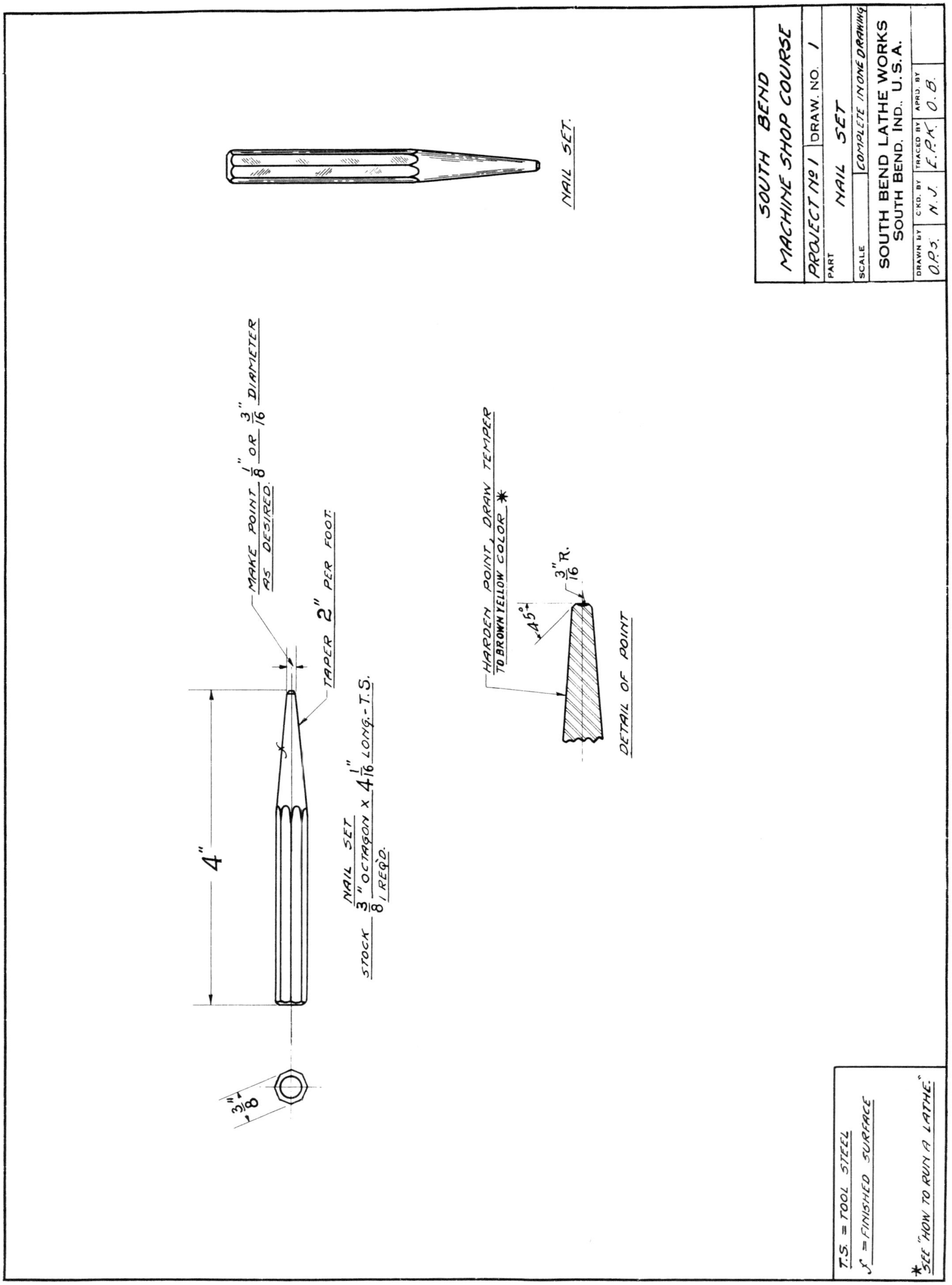
SOUTH BEND
MACHINE SHOP COURSE
PROJECT No 1
DRAW. NO. 1
PART
NAIL SET
SCALE
COMPLETE IN ONE DRAWING
SOUTH BEND LATHE WORKS
SOUTH BEND, IND. U.S.A.
DRAWN BY O.P.S.
C'KD. BY N.J.
TRACED BY E.P.K.
APR'D. BY O.B.
NAIL SET.
MAKE POINT 1/8" OR 3/16" DIAMETER
AS DESIRED.
TAPER 2" PER FOOT.
HARDEN POINT, DRAW TEMPER
TO BROWN YELLOW COLOR *
3/16" R.
45°
DETAIL OF POINT
4"
3/8"
NAIL SET
STOCK 3/8" OCTAGON x 4 1/16" LONG.-T.S.
1 REQ'D.
T.S. = TOOL STEEL
f = FINISHED SURFACE
* SEE "HOW TO RUN A LATHE."

NAIL SET

MATERIAL REQUIRED:

Tool steel 3/8″ octagon x 4 1/16″ long. One required.

OPERATIONS:

1. Select stock as per drawing.
2. Place steel in 4-jaw independent chuck, with end projecting 2 inches. Adjust jaws so stock runs true.*
3. Arrange belt for proper spindle speed.*
4. Set lathe tool for facing.*
5. Face end.*
6. Concave end. Use round nose tool ground to 3/16″ radius, setting tool on center.
7. Set compound rest at proper angle to obtain taper of 2 inches per foot. This is approximately 95 degrees to the right.
8. Set round nose tool on center for taper turning.*
9. Rough turn taper to approximate size*, feeding toward the headstock with compound rest screw.*
10. Finish turn taper using fine feed.*
11. Set tool on center for chamfering.
12. Chamfer point as per drawing.
13. File tapered section lightly to remove tool marks.*
14. Polish tapered section with emery cloth and oil.*
15. Turn stock end for end in chuck, allowing end to project 1 inch, and true as before.
16. Face end.
17. Chamfer corners.
18. Transfer to forge. Harden and temper taper end*, for distance of 1 inch from end, as per drawing.
19. Return to lathe and re-polish.

*See book "How to Run a Lathe."

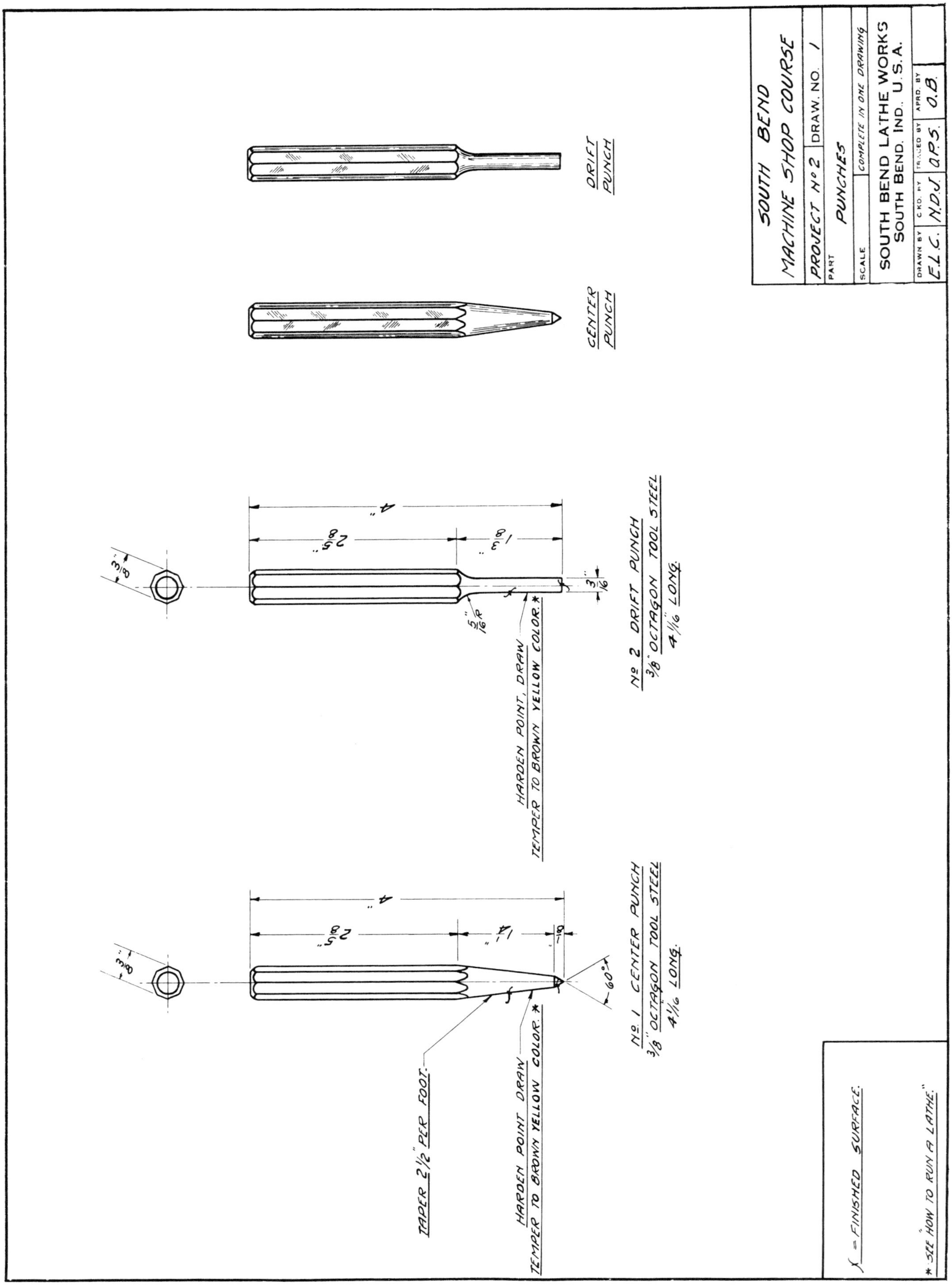
SOUTH BEND
MACHINE SHOP COURSE
PROJECT No 2
DRAW. NO. 1
PART
PUNCHES
SCALE
COMPLETE IN ONE DRAWING
SOUTH BEND LATHE WORKS
SOUTH BEND, IND., U.S.A.
DRAWN BY E.L.C.
CKD. BY N.D.J.
TRACED BY O.P.S.
APRD. BY O.B.
DRIFT PUNCH
CENTER PUNCH
No 2 DRIFT PUNCH
3/8" OCTAGON TOOL STEEL
4 1/16" LONG.
HARDEN POINT, DRAW
TEMPER TO BROWN YELLOW COLOR.*
No 1 CENTER PUNCH
3/8" OCTAGON TOOL STEEL
4 1/16" LONG.
TAPER 2 1/2" PER FOOT.
HARDEN POINT DRAW
TEMPER TO BROWN YELLOW COLOR.*
f = FINISHED SURFACE.
* SEE "HOW TO RUN A LATHE".

A SET OF PUNCHES

MATERIAL REQUIRED:

Tool steel 3/8″ octagon x 4 1/16″ long. Two required.

PART No. 1—CENTER PUNCH

OPERATIONS:

1. Select stock as per drawing.
2. Place stock in 4-jaw independent chuck, projecting 2 inches. Adjust jaws so stock runs true.*
3. Arrange belt for proper spindle speed.*
4. Set compound rest at proper angle to obtain taper of 2 1/2 inches per foot. This is approximately 96 degrees to the right.
5. Set tool for taper turning.*
6. Rough turn taper, using round nose tool and feeding toward headstock with compound rest screw.*
7. Finish turn taper using fine feed for smooth finish.*
8. Set compound rest 120 degrees to the right. Turn angle of point using same tool.
9. File machined section lightly to remove tool marks.*
10. Polish with fine emery cloth and oil.*
11. Turn stock end for end in chuck, projecting 1 inch, and true as before.
12. Set lathe tool for facing.*
13. Face end to length as per drawing.*
14. Set tool for chamfering.
15. Chamfer corner.
16. Transfer to forge, harden and temper* for 1/2 inch from point, as per drawing.
17. Replace in lathe chuck and re-polish.

PART No. 2—DRIFT PUNCH

1. Select stock as per drawing.
2. Place stock in 4-jaw independent chuck, projecting 2 inches. Adjust jaws so stock runs true.*
3. Arrange belt for proper spindle speed.*
4. Set lathe tool for facing.*
5. Face end.*
6. Set lathe tool for turning.*
7. Rough turn punch end.*
8. Finish turn punch end using round nose tool ground to form fillet at shoulder.
9. File machined section lightly to remove tool marks.*
10. Polish with fine emery cloth and oil.*
11. Turn stock end for end in chuck, projecting 1 inch, and true as before.
12. Set lathe tool for facing.
13. Face end to length as per drawing.
14. Set tool for chamfering.
15. Chamfer corners.
16. Transfer to forge. Harden and temper* for 1 3/4″ back from punch end.
17. Return to lathe chuck and re-polish.

*See book "How to Run a Lathe."

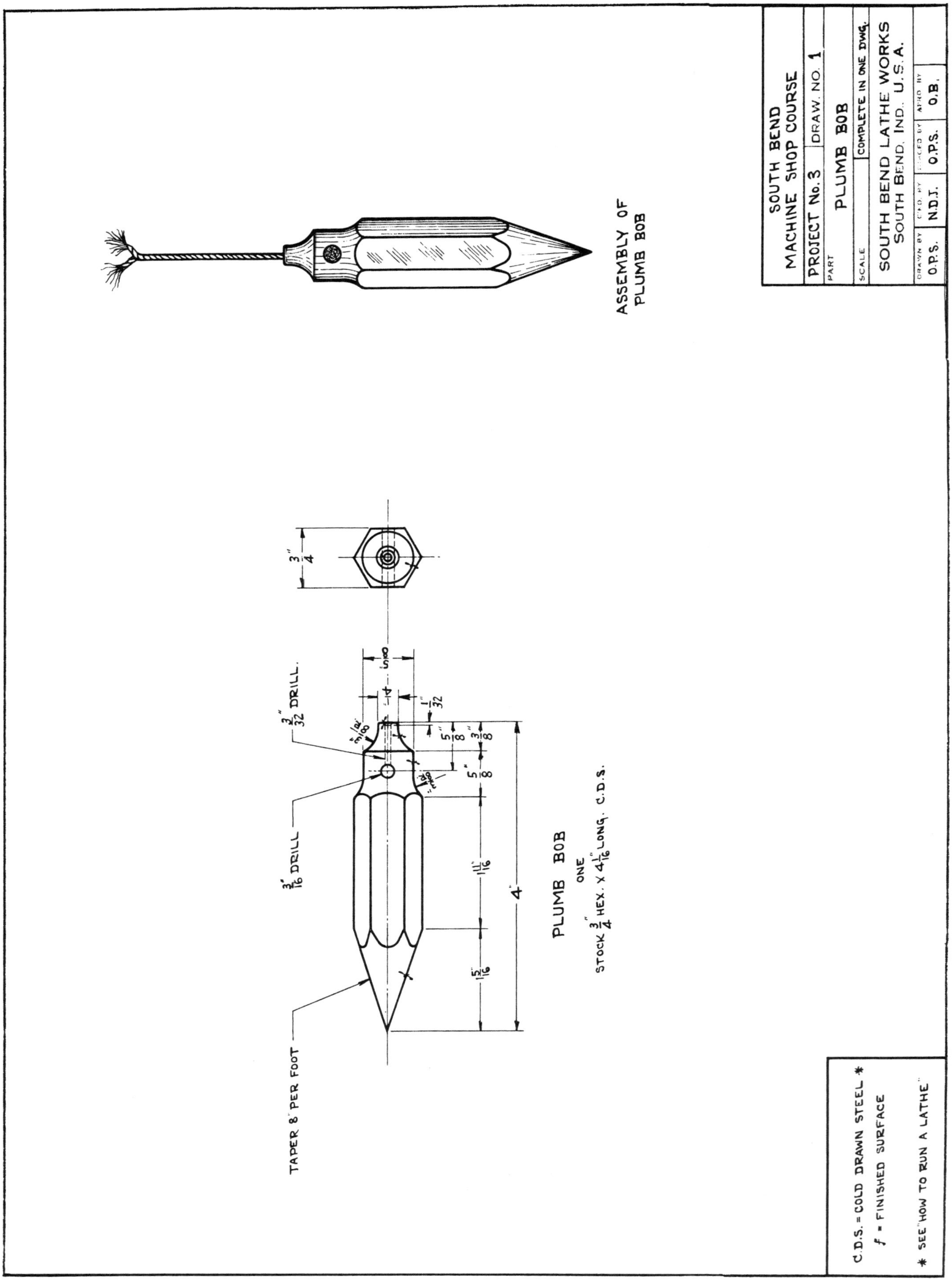
SOUTH BEND
MACHINE SHOP COURSE
PROJECT No. 3
DRAW. NO. 1
PART
PLUMB BOB
SCALE
COMPLETE IN ONE DWG.
SOUTH BEND LATHE WORKS
SOUTH BEND, IND. U.S.A.
DRAWN BY O.P.S.
CKD. BY N.D.I.
TRACED BY O.P.S.
APPD. BY O.B.
ASSEMBLY OF PLUMB BOB
3/4"
3/32" DRILL
3/16" DRILL
1/32"
5/8"
3/8"
5/8"
11/16
15/16
4"
TAPER 8" PER FOOT
PLUMB BOB
ONE
STOCK 3/4" HEX. X 4 1/16" LONG. C.D.S.
C.D.S. = COLD DRAWN STEEL *
f = FINISHED SURFACE
* SEE "HOW TO RUN A LATHE"

PLUMB BOB

MATERIAL REQUIRED:

Cold drawn steel ¾″ hex. x 4¼″ long. One required.

OPERATIONS:

1. Select stock as per drawing.
2. Lay out and center punch the 3/16″ hole near the end of stock as per drawing. In locating this hole allow for 1/32″ to be faced off at end of stock.
3. Transfer work to drill press. Drill hole through center of diameter as per drawing. (See footnote A.)
4. Transfer work to lathe. Place stock in a 3-jaw universal chuck with drilled end extending 1½″. Adjust so stock runs true.*
5. Arrange belt for proper spindle speed.*
6. Set lathe tool for facing.*
7. Face end.*
8. Center end of stock with sharp pointed tool held in tool post.
9. Center drill end as per drawing, with center drill held in drill chuck in tailstock spindle. Use center drill with 3/32″ pilot drill. (See footnote B.)
10. Drill 3/32″ hole in end of stock to intersect 3/16″ hole through diameter as per drawing.
11. Remove drill chuck from tailstock spindle and insert hardened center in its place.*
12. Adjust tailstock so it will support end of work.*
13. Set lathe tool for turning, using a round nose tool ground to form fillet at end of cut as per drawing.*
14. Rough turn large diameter section.*
15. Rough turn small diameter section.
16. Finish turn large diameter section.*
17. Finish turn small diameter section.
18. Round corner at end with file.
19. Polish machined surface with emery cloth.*
20. Turn stock end for end in chuck with 1¾″ extending. True as before.
21. Set compound rest at an angle of 108° (reading from headstock side) with handle toward tailstock.
22. Set lathe tool for taper turning.*
23. Rough turn point.
24. Finish turn point.
25. File tapered section* lightly to remove tool marks. Polish with fine emery cloth.*

NOTE A: Drilling can be done in the lathe using drill pad in tailstock spindle, and drill held in drill chuck in headstock spindle.*

NOTE B: Be sure that the center drill starts true so the center hole will be true to the axis of headstock spindle.

*See book "How to Run a Lathe."

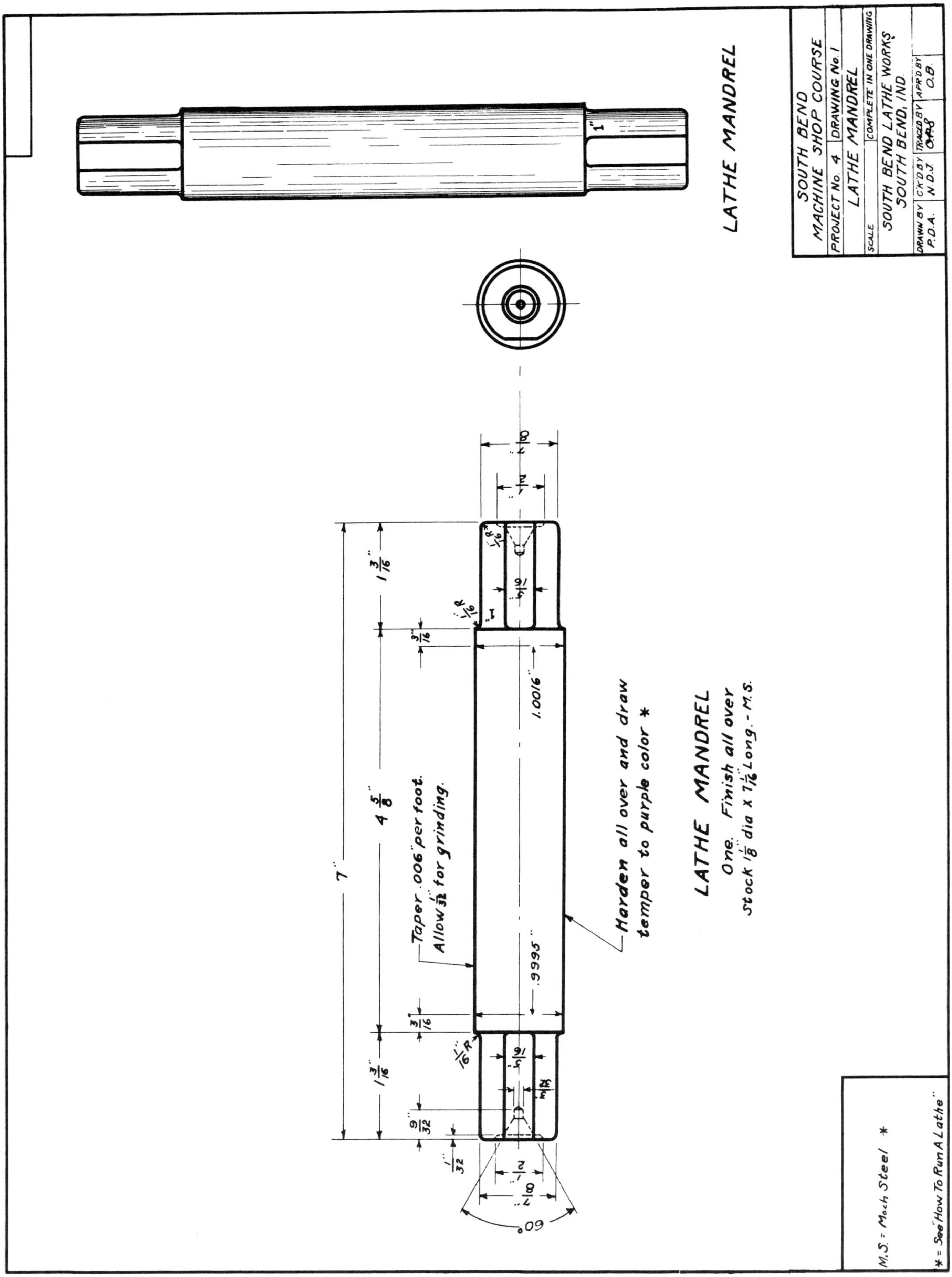
LATHE MANDREL
SOUTH BEND
MACHINE SHOP COURSE
PROJECT No. 4
DRAWING No. 1
LATHE MANDREL
SCALE
COMPLETE IN ONE DRAWING
SOUTH BEND LATHE WORKS
SOUTH BEND, IND.
DRAWN BY P.D.A.
CK'D BY N.D.J.
TRACED BY
APPR'D BY O.B.
1"
7"
4 5/8"
1 3/16"
3/16"
1.0016"
.9995"
9/32"
1/32"
1/16" R
60°
Taper .006" per foot.
Allow 1/32" for grinding.
Harden all over and draw
temper to purple color *
LATHE MANDREL
One. Finish all over
Stock 1 1/8" dia x 7 1/16" Long. - M.S.
M.S. = Mach. Steel *
* = See "How To Run A Lathe"

LATHE MANDREL

MATERIAL REQUIRED:

Machine steel 1⅛″ dia. x 7 1/16″ long. One required.

OPERATIONS:

1. Select stock as per drawing.
2. Lay off and center drill ends.*
3. Place stock on centers in lathe.* Drive with common lathe dog.* (See footnote A.)
4. Arrange belt for proper spindle speed.*
5. Set lathe tool for facing.*
6. Face end.*
7. Turn stock end for end on centers.
8. Face end to length as per drawing.
9. Set sharp pointed tool to undercut end around center hole.
10. Undercut end around center hole as per drawing to prevent marring center hole.
11. Turn stock end for end on centers.
12. Undercut end around hole on second end.
13. Check depth of center holes to drawing and re-center if necessary.
14. Place stock back on centers in lathe, using regular hardened tailstock center* in place of relieved center, and driving as before.
15. Set lathe tool for turning.*
16. Rough turn small diameter section on one end as per drawing.* Use round nose tool.*
17. Turn stock end for end on centers.
18. Rough turn small diameter section on second end.
19. Rough turn center section as per drawing, plus 1/32 inch on diameter for finishing.
20. Finish turn* small diameter section one end using round nose tool to form fillet* at shoulder.
21. Turn stock end for end on centers.
22. Finish turn small diameter section on second end.
23. Round end with forming tool as per drawing.
24. Turn stock end for end on centers. Place lathe dog so set screw is on line with set screw mark on other end of arbor.
25. Round second end.
26. Transfer to bench and file flat surfaces on small diameter sections as per drawing. File surfaces on proper location to remove set screw marks made by the dog.
27. Stamp size of arbor on small diameter section at one end of mandrel. (See footnote B.)
28. Transfer to forge and caseharden and temper* as per drawing. (See footnote C.)
29. Place mandrel on centers in lathe.
30. Set tailstock over to obtain taper of .006″ per foot, which is approximately .0023″ in length of center section of the mandrel.*
31. Take light trial cut over center section of mandrel.
32. Take measurements at both ends of center section with a micrometer caliper to test accuracy of taper.
33. If taper is not correct adjust setting to correct error. Take second trial cut and test as before. Repeat until taper is correct.
34. Finish turn (or grind) center section as per drawing. If finish turned, allow .003″ for filing.
35. File lightly to remove tool marks and obtain correct size.*
36. Polish with fine emery cloth and oil.*

NOTE A: A relieved center would simplify the facing operation.*

NOTE B: When finish grinding or turning mandrel see that size stamp is at large end of same.

NOTE C: If center section of mandrel is to be finished by grinding, harden all over. If to be turned, harden ends only for 1″ back.*

*See book "How to Run a Lathe."

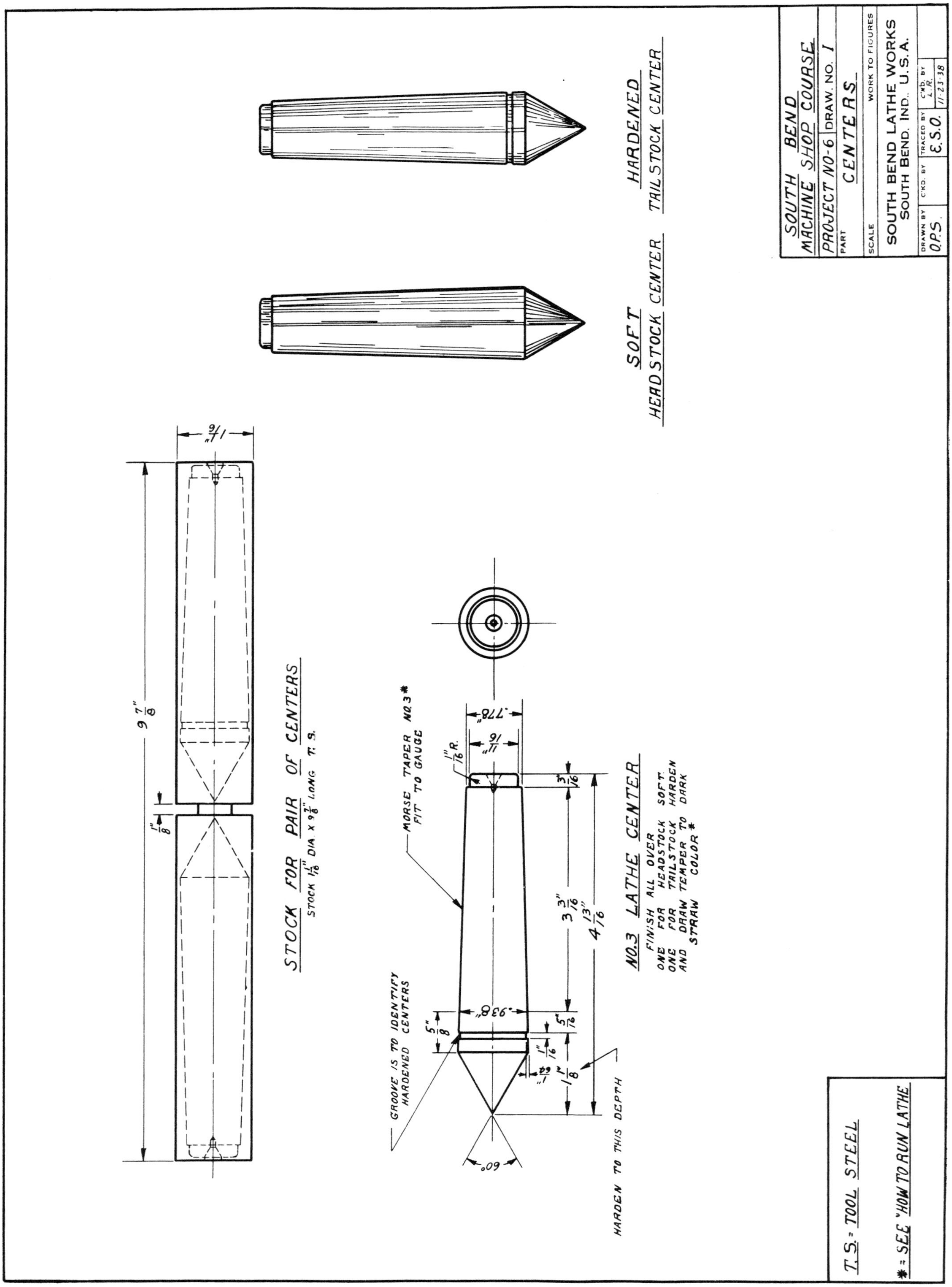
SOUTH BEND MACHINE SHOP COURSE
PROJECT NO-6
DRAW. NO. 1
PART CENTERS
SCALE WORK TO FIGURES
SOUTH BEND LATHE WORKS
SOUTH BEND, IND., U.S.A.
DRAWN BY O.P.S.
CKD. BY
TRACED BY E.S.O.
CKD. BY L.R. 11-23-38
HARDENED
TAILSTOCK CENTER
SOFT
HEADSTOCK CENTER
STOCK FOR PAIR OF CENTERS
STOCK 1 1/16" DIA x 9 7/8" LONG T.S.
1 1/16"
9 7/8"
1/8"
MORSE TAPER NO.3 * FIT TO GAUGE
.778"
11/16"
1/16" R.
3/16"
3 3/16"
4 13/16"
.938"
5/16"
5/8"
1/16"
1/64"
1 1/8"
60°
GROOVE IS TO IDENTIFY HARDENED CENTERS
HARDEN TO THIS DEPTH
NO.3 LATHE CENTER
FINISH ALL OVER
ONE FOR HEADSTOCK SOFT
ONE FOR TAILSTOCK HARDEN
AND DRAW TEMPER TO DARK
STRAW COLOR *
T.S. = TOOL STEEL
* = SEE "HOW TO RUN LATHE"

60° LATHE CENTERS, No. 3 MORSE TAPER

MATERIAL REQUIRED:

Tool steel 1 1/16" dia. x 9 7/8" long. One required.

OPERATIONS:

1. Select stock as per drawing.
2. Lay off and center ends.*
3. Place on centers in lathe; drive with common lathe dog.* (See footnote A.)
4. Arrange belt for proper spindle speed.*
5. Set lathe tool for facing.*
6. Face end.*
7. Turn stock end for end between centers.
8. Face end to length as per drawing.
9. Arrange lathe for taper turning.*
10. Set lathe tool for taper turning.*
11. Take a light trial cut over tapered section.
12. Test the taper in a No. 3 Morse Taper hole.*
13. If taper is not correct, adjust setting to correct error.
14. Take second cut and test as before. Repeat until you have taper correct.
15. Rough turn taper, allowing .020" stock for finishing cut.*
16. Turn stock end for end between centers.
17. Rough turn taper on other end allowing stock for finish as before.
18. Finish turn tapers to diameter, (as per drawing) on both ends, allowing .003" for filing. (See footnote B.) Soft brass should be placed around small end of taper to prevent marring of finished surface when lathe dog is attached.*

    NOTE: When taper shank is finished and inserted in taper gauge there should be a space of 5/8" between end of gauge and beginning of 60° angle on center point. See drawing.
19. Rearrange lathe for straight turning.*
20. Undercut both ends for clearance at small end of taper as per drawing. Round corners with file.
21. Cut ring mark on one center as per drawing.
22. File tapered section lightly to correct any slight error and to remove tool marks.*
23. Set 1/8" tool for cutting off.
24. Lay off and mark location of 1/8" recess in the middle of the stock.
25. Remove face plate and live center from headstock spindle.* Place taper shank in spindle and cut centers in two.*
26. Set compound rest at an angle of 60 degrees to the left.
27. Set lathe tool for taper turning.*
28. Turn the angle on center point which is 30° from center, both parts.
29. Test angle of point with thread center gauge.*
30. Harden and temper tailstock center (with ring mark).*
31. Polish tailstock center with emery cloth.*

    NOTE A: A relieved center would simplify the facing operation.*

    NOTE B: If it is desired to finish taper by grinding, allow .010" for finishing.

*See book "How to Run a Lathe."

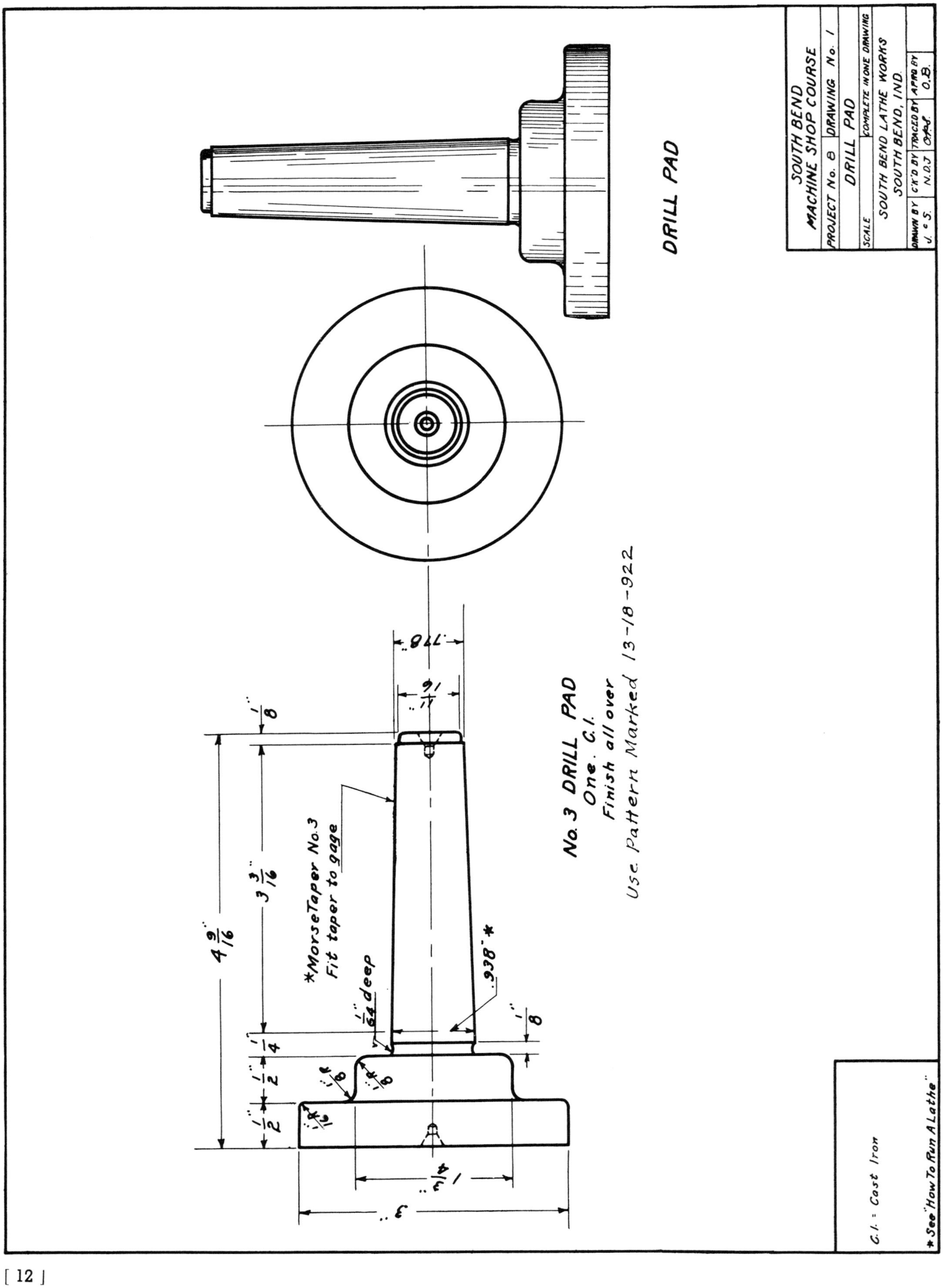
SOUTH BEND MACHINE SHOP COURSE
PROJECT No. 8
DRAWING No. 1
DRILL PAD
SCALE
COMPLETE IN ONE DRAWING
SOUTH BEND LATHE WORKS
SOUTH BEND, IND.
DRAWN BY J. S.
CK'D BY N.D.J.
APP'D BY O.B.
DRILL PAD
No. 3 DRILL PAD
One. C.I.
Finish all over
Use Pattern Marked 13-18-922
*Morse Taper No. 3
Fit taper to gage
C.I. = Cast Iron
* See "How To Run A Lathe"

DRILL PAD

MATERIAL REQUIRED:

Cast iron. One required.

OPERATIONS:

1. Secure casting.
2. Lay off and center ends.*
3. Place on centers in lathe.* Drive with clamp lathe dog at large end.* (See footnote A.)
4. Arrange belt for proper spindle speed.*
5. Set lathe tool for facing.*
6. Rough face small end.*
7. Finish face small end.
8. Turn stock end for end between centers driving with common lathe dog at small end.
9. Rough face large end.
10. Set lathe tool for turning.*
11. Rough turn outside diameter of flange.*
12. Rough face back of flange.* Use round nose tool.*
13. Rough turn hub. Use same tool.
14. Rough face hub.
15. Turn stock end for end between centers driving with clamp lathe dog at large end.*
16. Arrange lathe for taper turning.*
17. Set lathe tool for taper turning.*
18. Take light trial cut over tapered section.
19. Test the taper in No. 3 Morse taper socket.*
20. If taper is not correct, adjust setting to correct error.
21. Take second cut and test as before. Repeat until taper is correct.
22. Finish turn* tapered section, allowing .003″ for filing to diameter as per drawing. (See footnote B.)

    NOTE: When taper shank is finished and inserted in taper gauge there should be a space of ¼″ between end of gauge and hub of drill pad.
23. Arrange lathe for straight turning.*
24. Finish face hub.
25. Cut shoulder for clearance at small end of taper as per drawing. Round corner with file.*
26. File tapered section lightly to correct any slight error and to remove tool marks.*
27. Remove face plate and live center from spindle nose* and place taper shank in lathe spindle.
28. Finish turn flange.
29. Finish turn hub.
30. Finish face back of flange. Use round nose tool ground to form fillet in corner.*
31. Round corner of hub with tool, and file.
32. Finish face front of flange.
33. File flange and hub lightly to remove tool marks. Polish with emery cloth.*

    NOTE A: A relieved center would simplify the face operation.*

    NOTE B: If it is desired to finish taper by grinding, finish turn to .010″ oversize and undercut at large end as per drawing for clearance for grinding wheel.

*See book "How to Run a Lathe."

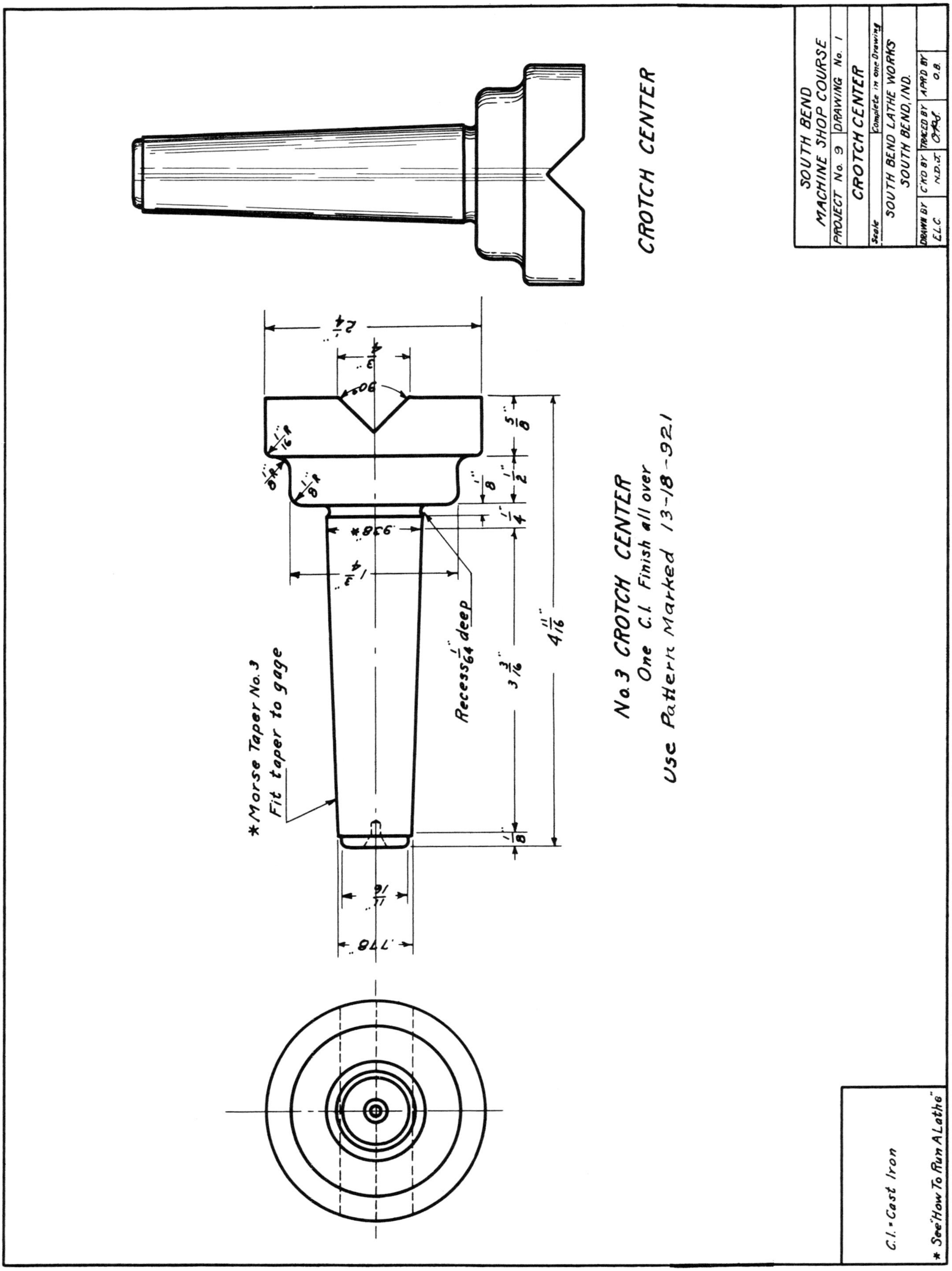
CROTCH CENTER
SOUTH BEND
MACHINE SHOP COURSE
PROJECT No. 9
DRAWING No. 1
CROTCH CENTER
Scale
Complete in one Drawing
SOUTH BEND LATHE WORKS
SOUTH BEND, IND.
DRAWN BY E.L.C.
C'KD BY N.D.J.
APP'D BY O.B.
No.3 CROTCH CENTER
One C.I. Finish all over
Use Pattern Marked 13-18-921
*Morse Taper No.3
Fit taper to gage
Recess 1/64" deep
2 1/4"
3/4"
90°
5/8"
1/2"
1/8"
1/4"
3 3/16"
4 11/16"
.938*
1 3/4"
11/16"
.778"
1/16 R
1/8 R
C.I. = Cast Iron
* See "How To Run A Lathe"

CROTCH CENTER

MATERIAL REQUIRED:

Cast iron. One required.

OPERATIONS:

1. Secure casting.
2. Lay off and center ends.*
3. Place on centers in lathe,* drive with clamp lathe dog at large end.* (See footnote A.)
4. Arrange belt for proper spindle speed.*
5. Set lathe tool for facing.*
6. Rough face small end.*
7. Finish face small end.
8. Turn stock end for end between centers driving with common lathe dog at small end.
9. Rough face large end.
10. Set lathe tool for turning.*
11. Rough turn outside diameter of flange.*
12. Rough face back of flange.* Use round nose tool.
13. Rough turn hub. Use same tool.
14. Rough face hub.
15. Turn stock end for end between centers driving with clamp lathe dog at large end.*
16. Arrange lathe for taper turning.*
17. Set lathe tool for taper turning.*
18. Take light trial cut over tapered section.
19. Test the taper in a No. 3 Morse taper socket.*
20. If taper is not correct, adjust setting to correct error.
21. Take second cut and test as before. Repeat until taper is correct.
22. Finish turn* tapered section, allowing .003″ for filing to diameter as per drawing. (See footnote B.)

    NOTE: When taper shank is finished and inserted in taper gauge there should be a space of ¼″ between end of gauge and hub of crotch center.
23. Arrange lathe for straight turning.*
24. Finish face hub.
25. Cut shoulder for clearance at small end of taper as per drawing. Round corner with file.*
26. File tapered section lightly to correct any slight error and to remove tool marks.*
27. Remove face plate and live center from spindle nose* and place taper shank in lathe spindle.
28. Finish turn flange.
29. Finish turn hub.
30. Finish face back of flange. Use tool ground to form fillet in corner.*
31. Round corner of hub with tool, and file.
32. Finish face front of flange.
33. File flange and hub lightly to remove tool marks. Polish with emery cloth.*
34. Transfer to milling machine. Mill crotch.

    NOTE A: A relieved center would simplify the facing operation.*

    NOTE B: If it is desired to finish taper by grinding, finish turn to .010″ oversize and undercut at large end as per drawing for clearance for grinding wheel.

*See book "How to Run a Lathe."

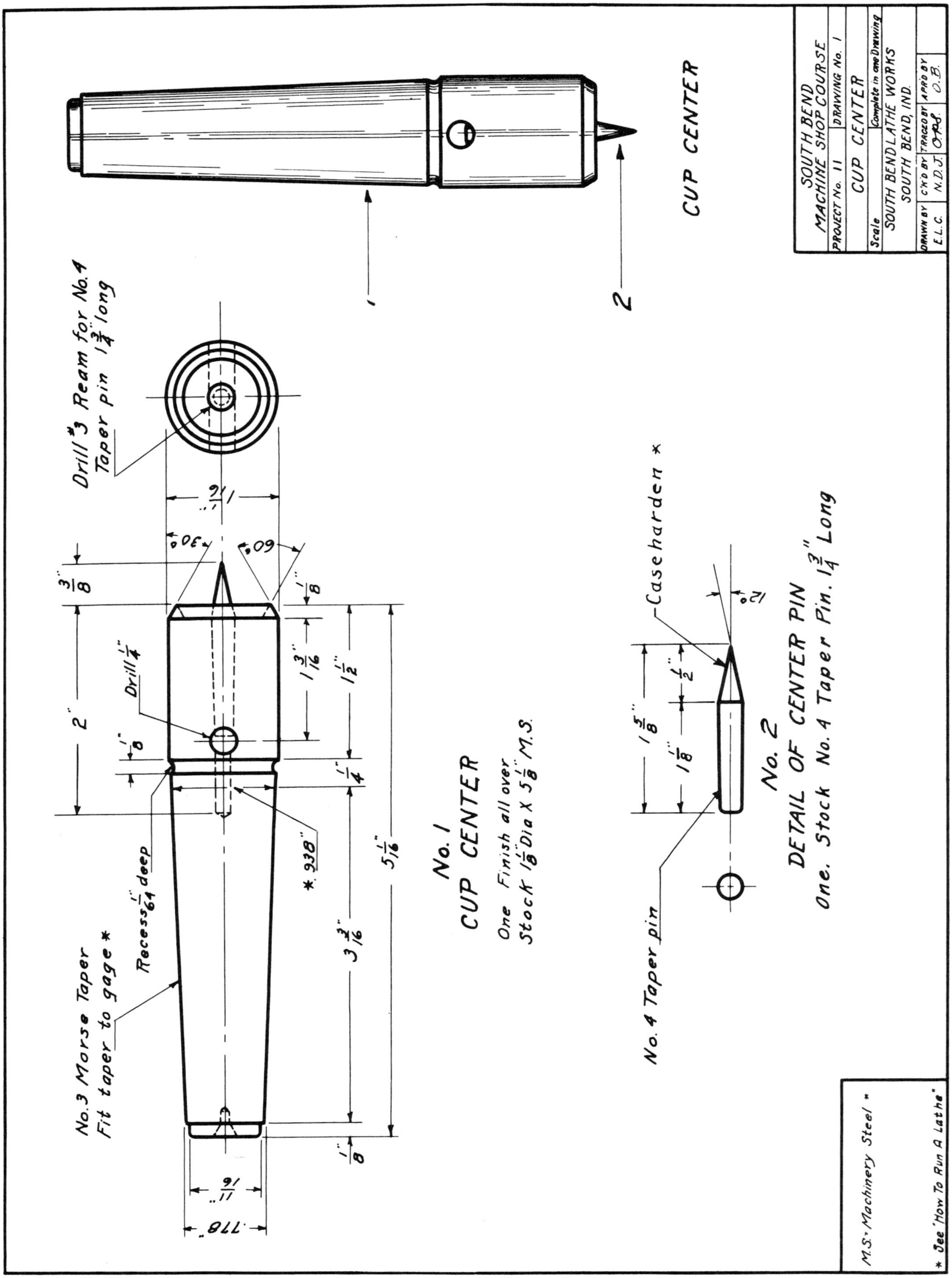
SOUTH BEND
MACHINE SHOP COURSE
PROJECT No. 11
DRAWING No. 1
CUP CENTER
Scale
Complete in one Drawing
SOUTH BEND LATHE WORKS
SOUTH BEND, IND.
DRAWN BY
C'K'D BY
TRACED BY
APP'D BY
CUP CENTER
Drill #3 Ream for No. 4 Taper pin 1 3/4" long
Drill 1/4"
No. 3 Morse Taper
Fit taper to gage *
Recess 1/64" deep
No. 1
CUP CENTER
One Finish all over
Stock 1 1/8" Dia X 5 1/8" M.S.
Caseharden *
No. 4 Taper pin
No. 2
DETAIL OF CENTER PIN
One. Stock No. 4 Taper Pin. 1 3/4" Long
M.S.-Machinery Steel *
* See 'How To Run A Lathe'

CUP CENTER

MATERIAL REQUIRED:

Part No. 1—Machinery steel 1⅛″ dia. x 5⅛″ long. One required.
Part Nos. 2 and 3—Standard No. 4 taper pin 1¾″ long. Two required.

OPERATIONS:

1. Select stock as per drawing.
2. Lay off and center ends.*
3. Place stock on centers in lathe.* Drive with common lathe dog.* (See footnote A.)
4. Arrange belt for proper spindle speed.*
5. Set lathe tool for facing.*
6. Face end.*
7. Turn stock end for end on centers.
8. Face end to length as per drawing.
9. Set lathe tool for turning.*
10. Rough turn straight diameter section.*
11. Turn stock end for end on centers.
12. Arrange lathe for taper turning.*
13. Set lathe tool for taper turning.*
14. Take light trial cut over tapered section.
15. Test the taper in No. 3 Morse taper socket.*
16. If taper is not correct, adjust setting to correct error.
17. Take second cut and test as before. Repeat until taper is correct.
18. Rough turn taper allowing .020″ for finishing cut.
19. Finish turn* tapered section, allowing .003″ for filing to diameter as per drawing. (See footnote B.)
    NOTE: When taper shank is finished and inserted in taper gauge, there should be a space of ¼″ between end of gauge and straight section of spur center.
20. Arrange lathe for straight turning.*
21. Cut shoulder for clearance at small end of taper as per drawing. Round corner with file.*
22. File tapered section lightly to correct any slight error and to remove tool marks.
23. Remove face plate and live center from spindle nose* and place taper shank in lathe spindle.
24. Finish turn straight section.*
25. File and polish straight section.*
26. Transfer to bench. Lay off and center punch for hole through diameter as per drawing.
27. Transfer to drill press. Drill hole through diameter. (See footnote C.)
28. Transfer back to lathe. Insert taper shank in lathe spindle.
29. Drill hole lengthwise 2 inches deep and intersecting hole through diameter. Use No. 3 drill held in drill chuck in tailstock spindle.
30. Re-drill hole ⅝″ deep from face of cup center, with No. 1 drill to guide reamer.
31. Set compound rest at an angle of 60 degrees to the left.
32. Set small round nose tool for taper turning.*
33. Turn outside angle of cup feeding with compound rest screw.*
34. Set compound rest at an angle of 60 degrees to the right.
35. Set same tool for facing.
36. Rough bore inside of cup.
37. Set right-hand side facing tool, (slightly rounded at point) for taper turning.*
38. Finish face bottom of cup, feeding with cross-feed screw.
39. Finish turn inside angle of cup, feeding with compound rest screw.*
40. Insert No. 4 taper pin reamer into drill chuck in tailstock spindle.
41. Pull belt by hand, and feed reamer slowly into pin hole for short distance to start reamer true. Remove reamer from hole often and clean chips out of flutes of same to prevent breaking reamer.
42. Transfer to bench. Finish reaming hole for taper pin by hand. (See footnote D.)
43. Remove all dirt from taper pin hole and pin.
44. Oil pin. Place same in taper hole and drive to seat.
45. Replace taper shank in lathe spindle. Set compound rest at 102 degrees to the right.
46. Set tool for taper turning.*
47. Turn point on pin.* Use tool ground with long narrow point for clearance.
48. Remove center pin and caseharden for 9/16″ back from point.*
49. To remove center pin use the second No. 4 taper pin, (filed flat on one side) as a drift. This drift should be casehardened all over.

NOTE A: A relieved center would simplify the facing operation.*
NOTE B: If it is desired to finish taper by grinding, finish turn to .010″ oversize and undercut at large end as per drawing for clearance for grinding wheel.
NOTE C: Drilling can be done in lathe, using crotch center in tailstock spindle, and drill held in drill chuck in headstock spindle.*
NOTE D: This hole should be reamed so that when taper pin is properly seated the small end of same will project 1/16″ into hole through diameter.

*See book "How to Run a Lathe."

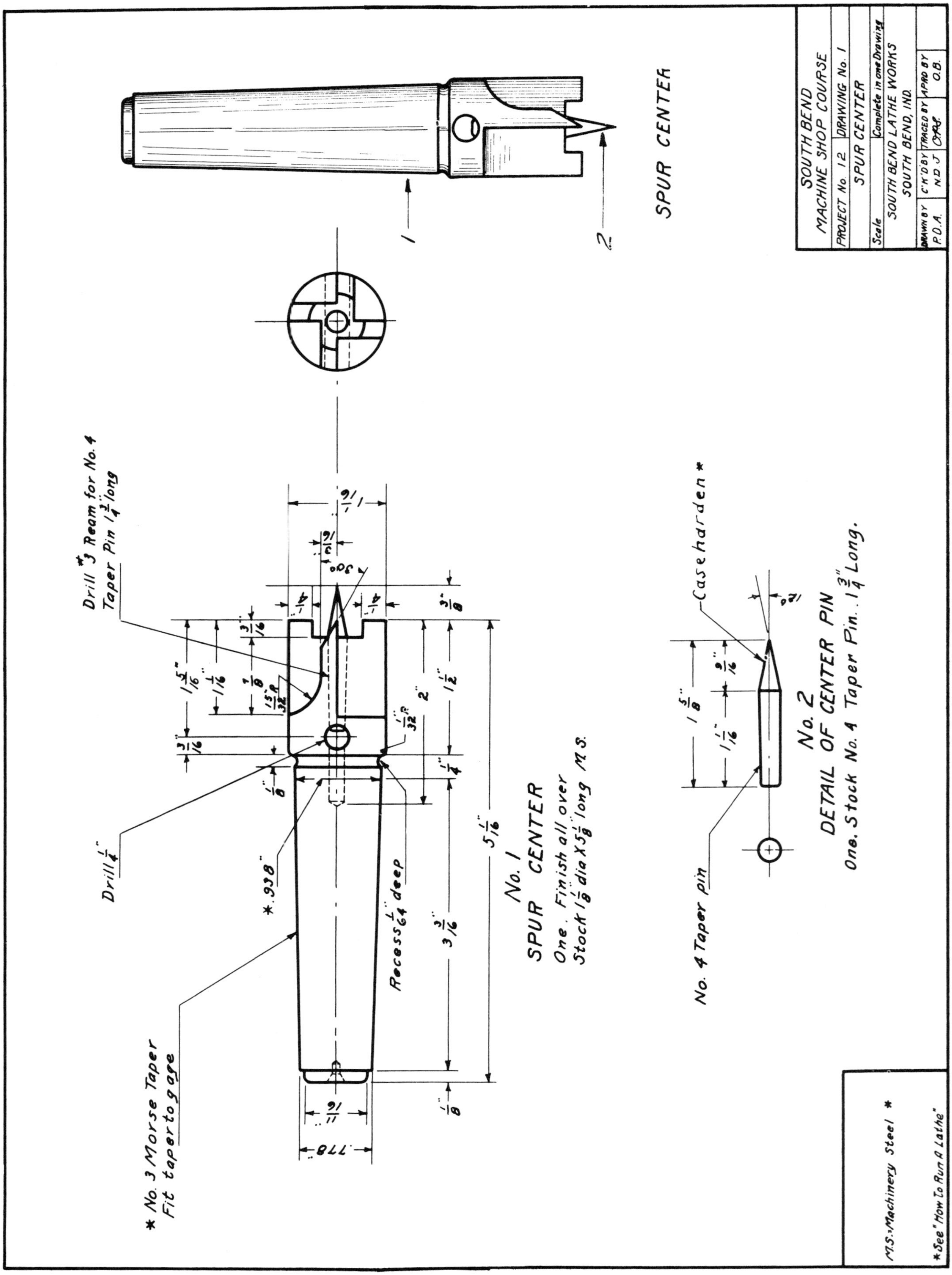
SOUTH BEND
MACHINE SHOP COURSE
PROJECT No. 12
DRAWING No. 1
SPUR CENTER
Scale
Complete in one Drawing
SOUTH BEND LATHE WORKS
SOUTH BEND, IND.
DRAWN BY P.O.A.
C'K'D BY N.D.J.
APP'D BY O.B.
SPUR CENTER
Drill #3 Ream for No. 4 Taper Pin 1 1/4" long
Drill 1/4"
*.938"
Recess 1/64" deep
* No. 3 Morse Taper Fit taper to gage
No. 1
SPUR CENTER
One. Finish all over
Stock 1 1/8" dia X 5 1/8" long M.S.
Casehardden *
No. 4 Taper pin
No. 2
DETAIL OF CENTER PIN
One. Stock No. 4 Taper Pin. 1 3/4" Long.
M.S.=Machinery Steel *
*See "How To Run A Lathe"

SPUR CENTER

MATERIAL REQUIRED:

Part No. 1—Machinery steel 1⅛″ dia. x 5⅛″ long. One required.
Part Nos. 2 and 3—Standard No. 4 taper pin 1¾″ long. Two required.

OPERATIONS:

1. Select stock as per drawing.
2. Lay off and center ends.*
3. Place on centers in lathe.* Drive with common lathe dog.* (See footnote A.)
4. Arrange belt for proper spindle speed.*
5. Set lathe tool for facing.*
6. Face end.*
7. Turn stock end for end on centers.
8. Face end to length as per drawing.
9. Set lathe tool for turning.*
10. Rough turn straight diameter section.*
11. Turn stock end for end on centers.
12. Arrange lathe for taper turning.*
13. Set lathe tool for taper turning.*
14. Take light trial cut over tapered section.
15. Test the taper in No. 3 Morse taper hole.*
16. If taper is not correct, adjust setting to correct error.
17. Take second cut and test as before. Repeat until taper is correct.
18. Rough turn taper* section, allowing .020″ for finishing cut.
19. Finish turn* tapered section, allowing .003″ for filing to diameter as per drawing. (See footnote B.)

    NOTE: When taper shank is finished and inserted in taper gauge, there should be a space of ¼″ between end of gauge and straight section of spur center.
20. Arrange lathe for straight turning.*
21. Cut shoulder for clearance at small end of taper as per drawing. Round corner with file.*
22. File tapered section lightly to correct any slight error and to remove tool marks.*
23. Remove face plate and live center from spindle nose* and place taper shank in lathe spindle.
24. Finish turn straight section.*

NOTE A: A relieved center would simplify the facing operation.*

NOTE B: If it is desired to finish taper by grinding, finish turn to .010″ oversize and undercut at large end as per drawing, for clearance for grinding wheel.

25. File and polish straight section.*
26. Transfer to bench. Lay off and center punch for hole through diameter as per drawing.
27. Transfer to drill press. Drill hole through diameter. (See footnote C.)
28. Transfer back to lathe. Insert tapered shank in lathe spindle.
29. Drill small hole lengthwise 2″ deep and intersecting hole through diameter. Use No. 3 drill held in drill chuck in tailstock spindle.*
30. Redrill hole 11/16″ deep from face of spur center with No. 1 drill in chuck, to guide reamer.
31. Counterbore face as per drawing. Use tool ground to long narrow point for clearance.
32. Insert No. 4 taper pin reamer in drill chuck. Put oil on reamer.
33. Pull belt by hand and feed reamer slowly into pin hole for short distance to start reamer true. Remove reamer from hole often and clean chips out of flutes of same to prevent breaking reamer.
34. Transfer to bench. Finish reaming hole for taper pin by hand. (See footnote D.)
35. Remove all dirt from taper pin hole and taper pin.
36. Oil pin. Place in taper hole and drive to seat.
37. Replace taper shank in lathe spindle. Set compound rest at 102 degrees to the right.
38. Set tool for taper turning.*
39. Turn point on pin.* Use tool ground with long narrow point for clearance.
40. Transfer to milling machine. Mill flutes.
41. Transfer to bench. File 30 degree angle on spur points as per drawing.
42. File off burrs and sharp corners.
43. Remove center pin and caseharden for 9/16″ back from point.*
44. To remove center pin use the second No. 4 taper pin, (filed flat on one side) as a drift. This drift should be casehardened all over.

NOTE C: Drilling can be done in lathe using crotch center in tailstock spindle, and drill held in drill chuck in headstock spindle.*

NOTE D: This hole should be reamed so that when taper pin is properly seated the small end of same will project 1/16″ into hole through diameter.

*See book "How to Run a Lathe."

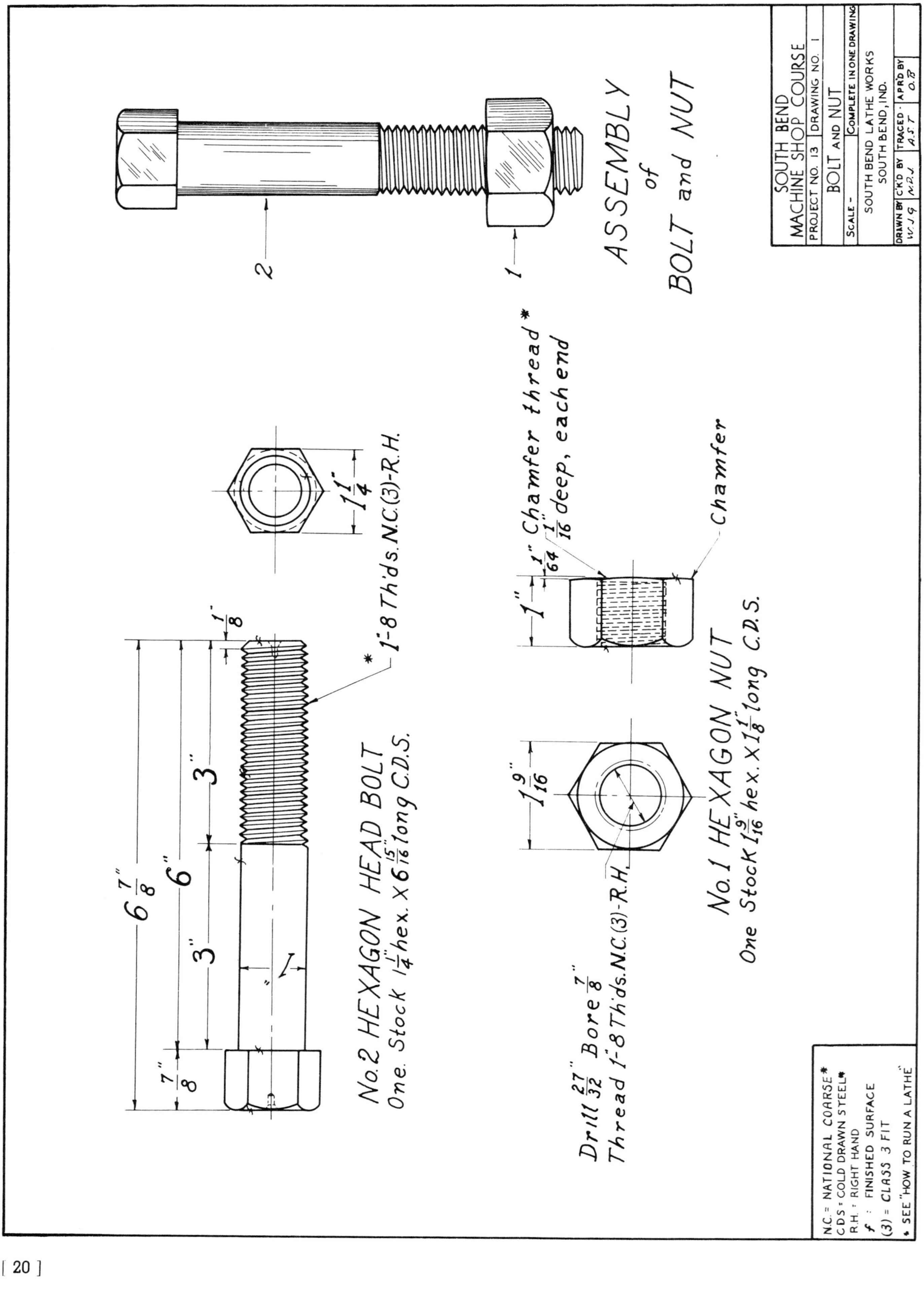
SOUTH BEND MACHINE SHOP COURSE
PROJECT NO. 13
DRAWING NO. 1
BOLT AND NUT
SCALE -
COMPLETE IN ONE DRAWING
SOUTH BEND LATHE WORKS SOUTH BEND, IND.
DRAWN BY W.J.G.
CK'D BY N.D.J.
TRACED A.S.T.
APR'D BY O.B.
ASSEMBLY of BOLT and NUT
2
1
No.2 HEXAGON HEAD BOLT
One Stock 1 1/4" hex. X 6 15/16" long C.D.S.
* 1"-8 Th'ds. N.C.(3)-R.H.
6 7/8"
6"
3"
3"
7/8"
1/8"
1 1/4"
No.1 HEXAGON NUT
One Stock 1 9/16" hex. X 1 1/8" long C.D.S.
Drill 27/32" Bore 7/8"
Thread 1"-8 Th'ds. N.C.(3)-R.H.
1/64" Chamfer thread*
1/16" deep, each end
Chamfer
1"
1 9/16
N.C. = NATIONAL COARSE*
CDS = COLD DRAWN STEEL*
R.H. = RIGHT HAND
f = FINISHED SURFACE
(3) = CLASS 3 FIT
* SEE "HOW TO RUN A LATHE"

1″ BOLT AND NUT

MATERIAL REQUIRED:

Part No. 1—Cold drawn steel—1 9/16″ hex. x 1 1/8″ long. One required.
Part No. 2—Cold drawn steel—1 1/4″ hex. x 6 15/16″ long. One required.

PART No. 1—HEXAGON NUT

OPERATIONS:

1. Select stock as per drawing.
2. Place stock in 3-jaw universal chuck with 1/4″ extending and running true.*
3. Arrange belt for proper spindle speed.*
4. Set lathe tool for facing.*
5. Face end.*
6. Center end of stock. Use sharp pointed tool held in tool post.
7. Rough drill hole through stock as per drawing with drill held in drill chuck in tailstock spindle.*
8. Set tool for boring.*
9. Bore hole to size as per drawing.
10. Arrange gears for cutting thread as per drawing.*
11. Set tool for thread cutting.*
12. Cut thread, leaving .005″ stock for tap to remove.*
13. Finish thread to size with tap, holding tap with wrench and guiding with tailstock center.*
14. Chamfer thread as per drawing.
15. Chamfer outside corners lightly as per drawing.
16. Turn stock end for end in chuck, truing as before.
17. Face end to length as per drawing.
18. Chamfer thread as per drawing.
19. Chamfer corners as per drawing. Set tool at an angle of approximately 45 degrees.

PART No. 2—HEXAGON HEAD BOLT

1. Select stock as per drawing.
2. Lay off and center ends.*
3. Place stock on centers in lathe;* drive with clamp lathe dog.* (See footnote A.)
4. Arrange belt for proper spindle speed.*
5. Set lathe tool for facing.*
6. Face end.*
7. Turn stock end for end between centers.
8. Face end to length as per drawing.
9. Chamfer corners. Set tool at an angle of approximately 45 degrees.
10. Turn stock end for end on centers.
11. Set lathe tool for turning.*
12. Rough turn round section.*
13. Rough face shoulder at end of round section. Use same tool.
14. Set lathe tool for facing.
15. Finish face shoulder as per drawing.
16. Set lathe tool for finish turning.
17. Finish turn round section as per drawing.*
18. Turn thread section (A) .010″ undersize to insure a free fit on thread.
19. Arrange gears for cutting thread as per drawing.*
20. Set tool for thread cutting.*
21. Cut thread as per drawing,* fitting to Part No. 1—Hexagon Nut.
22. Chamfer end of threaded section as per drawing. Set tool as before.

NOTE A: A relieved center would simplify the facing operation.

*See book "How to Run a Lathe."

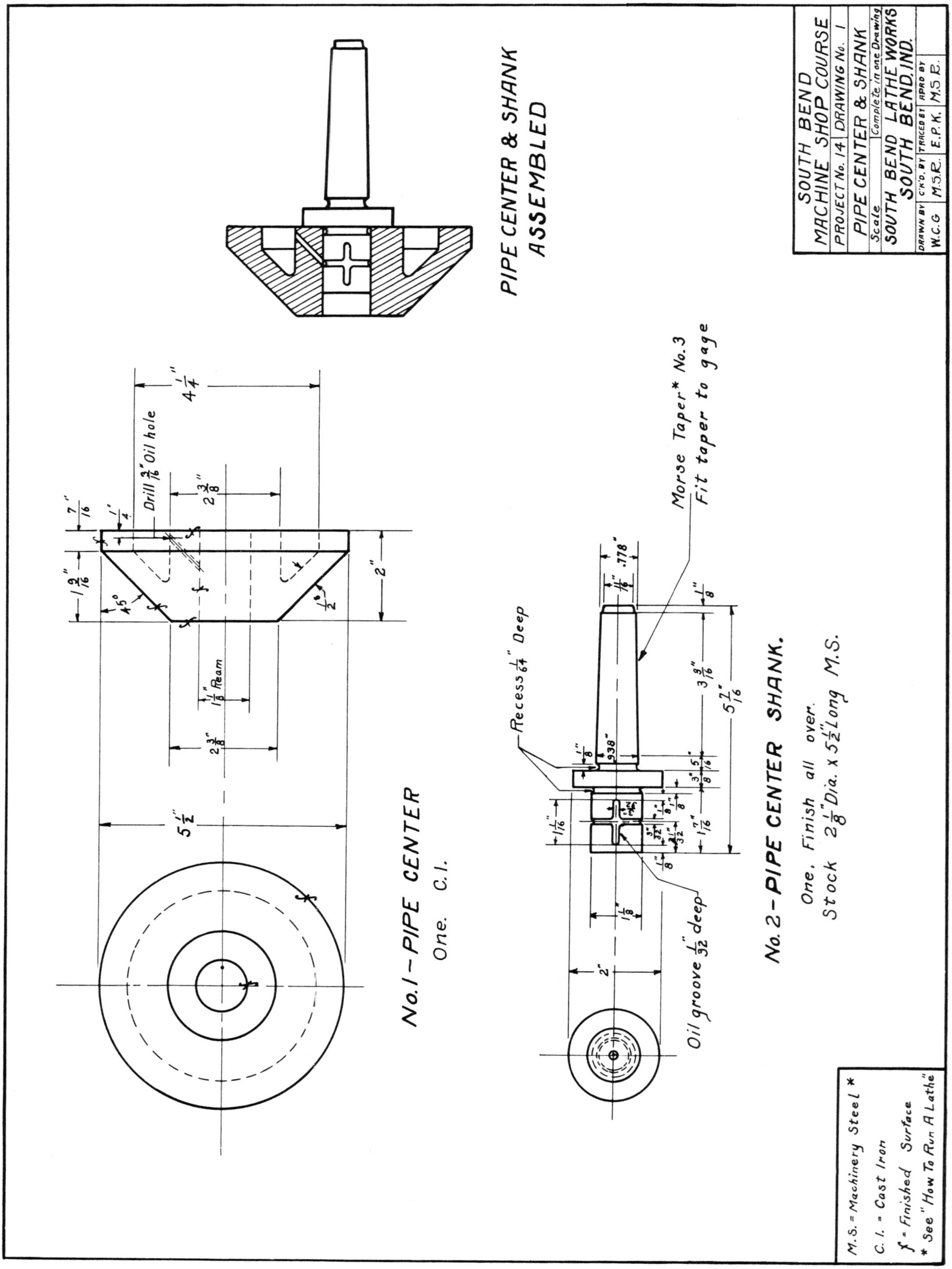
SOUTH BEND
MACHINE SHOP COURSE
PROJECT No. 14
DRAWING No. 1
PIPE CENTER & SHANK
Scale
Complete in one Drawing
SOUTH BEND LATHE WORKS
SOUTH BEND, IND.
DRAWN BY W.C.G
C'K'D BY M.S.R.
TRACED BY E.P.K.
APP'D BY M.S.R.
PIPE CENTER & SHANK
ASSEMBLED
Drill 3/16" Oil hole
1 1/8" Ream
No. 1 – PIPE CENTER
One. C.I.
Morse Taper* No. 3
Fit taper to gage
Recess 1/64" Deep
Oil groove 1/32" deep
No. 2 – PIPE CENTER SHANK.
One. Finish all over.
Stock 2 1/8" Dia. x 5 1/2" Long M.S.
M.S. = Machinery Steel *
C.I. = Cast Iron
f = Finished Surface
* See "How To Run A Lathe"

PIPE CENTER AND SHANK

MATERIAL REQUIRED:

Part No. 1—Cast iron. One required.
Part No. 2—Machinery steel 2⅛″ dia. x 5½″ long. One required.

OPERATIONS:

PART No. 1—PIPE CENTER

1. Remove sand and rough projections from casting.
2. Place casting in 4-jaw Independent Chuck, with straight side out and projecting ¼″. Adjust jaws so stock runs true.*
3. Arrange belt for proper spindle speed.*
4. Set round nose tool for facing.*
5. Rough face hub and flange of casting.
6. Center casting for drilling. Use sharp pointed tool held in tool post.*
7. Drill through casting with 1″ drill held in drill chuck in tailstock spindle.*
8. Set lathe tool for boring.*
9. Bore hole as per drawing. Allow .010″ for reaming hole to size.
10. Ream hole to size as per drawing. Hold machine reamer in drill chuck in place of drill.*
11. Set tool for facing.*
12. Finish face hub and flange of casting.*
13. Transfer work to arbor press. Press lathe mandrel in hole in casting. (See footnote A.)
14. Place mandrel on centers in lathe with tapered side of casting toward tailstock. Drive with common lathe dog.*
15. Set round nose tool for facing.*
16. Rough face unfinished side of casting.
17. Set side tool for facing.*
18. Finish face casting to width as per drawing.
19. Set round nose tool for turning.*
20. Rough turn large diameter section.
21. Set compound rest at an angle of 45 degrees to the left.
22. Set tool for taper turning.*
23. Rough turn angle, feeding with compound rest screw.*
24. Finish turn large diameter section, feeding with automatic longitudinal carriage feed.
25. Finish turn angle feeding as before. Use fine feed for smooth finish.
26. File machined surfaces to remove tool marks.* File corners lightly to break sharp edge. Polish with emery cloth.*
27. Center punch for the 3⁄16″ oil hole as per drawing.
28. Transfer work to drill press. Drill oil hole at approximately 45 degrees as per drawing.

PART No. 2—PIPE CENTER SHANK

1. Select stock as per drawing.
2. Lay off and center ends.*
3. Place stock on centers in lathe. Drive with common lathe dog. (See footnote B.)
4. Arrange belt for proper spindle speed.*
5. Set lathe tool for facing.*
6. Face end.*
7. Turn stock end for end on centers.
8. Face end to length as per drawing.
9. Set round nose tool for turning.*
10. Rough turn 1⅛″ diameter section.*
11. Rough face shoulder of collar. Use same tool.*
12. Rough turn diameter of collar.*
13. Turn stock end for end on centers.
14. Rough turn taper section to approximately 1⅓₂″ diameter.
15. Rough face shoulder of collar. Use same tool.
16. Arrange lathe for taper turning.*
17. Take light trial cut over tapered section.
18. Test the taper in a No. 3 Morse taper socket.*
19. If taper is not correct adjust setting to correct error.
20. Take a second cut and test as before. Repeat until taper is correct.
21. Rough turn taper, allowing .020″ stock for finishing cut.
22. Finish turn taper, allowing .003″ for filing to diameter as per drawing. (See footnote C.)

   NOTE: When taper shank is finished and inserted in taper gauge there should be a space of 5⁄16″ between end of gauge and shoulder at large end of taper. Allow for stock to finish face shoulder.
23. Rearrange lathe for straight turning.
24. Cut shoulder for clearance at small end of taper as per drawing. Round corner with file.*
25. File tapered section lightly to correct any slight error and to remove tool marks.* Polish with emery cloth.
26. Set lathe tool for facing.*
27. Finish face shoulder of collar.*
28. Turn work end for end on centers.
29. Set lathe tool for finish turning.
30. Finish turn 1⅛″ diameter section, allowing .003″ for filing to size as per drawing. (See footnote D.)
31. Finish turn diameter of collar section.
32. Set lathe tool for facing.
33. Finish face shoulder of collar to width as per drawing.
34. Turn oil groove in 1⅛″ diameter section as per drawing. Use small round nose tool.
35. File machined surfaces lightly to remove tool marks.* Polish with emery cloth.*
36. Transfer to shaper or milling machine and plane or mill oil groove running lengthwise of work. (See footnote E.)
37. Remove burr from sides of oil groove with a file.

NOTE A: Remove dirt and put oil in hole before pressing mandrel into casting. Also clean and oil the mandrel.*

NOTE B: A relieved center would simplify the facing operation.*

NOTE C: If it is desired to finish the taper by grinding, finish turn to .010″ overside and undercut at large end as per drawing for clearance for grinding wheel.

NOTE D: If it is desired to finish the 1⅛″ dia. by grinding, finish turn to .010″ oversize and cut recess next to shoulder for clearance for the grinding wheel.

NOTE E: If shaper or milling machine is not available, transfer work to vise and chip oil groove. Use small round nose chisel.

*See book "How to Run a Lathe."

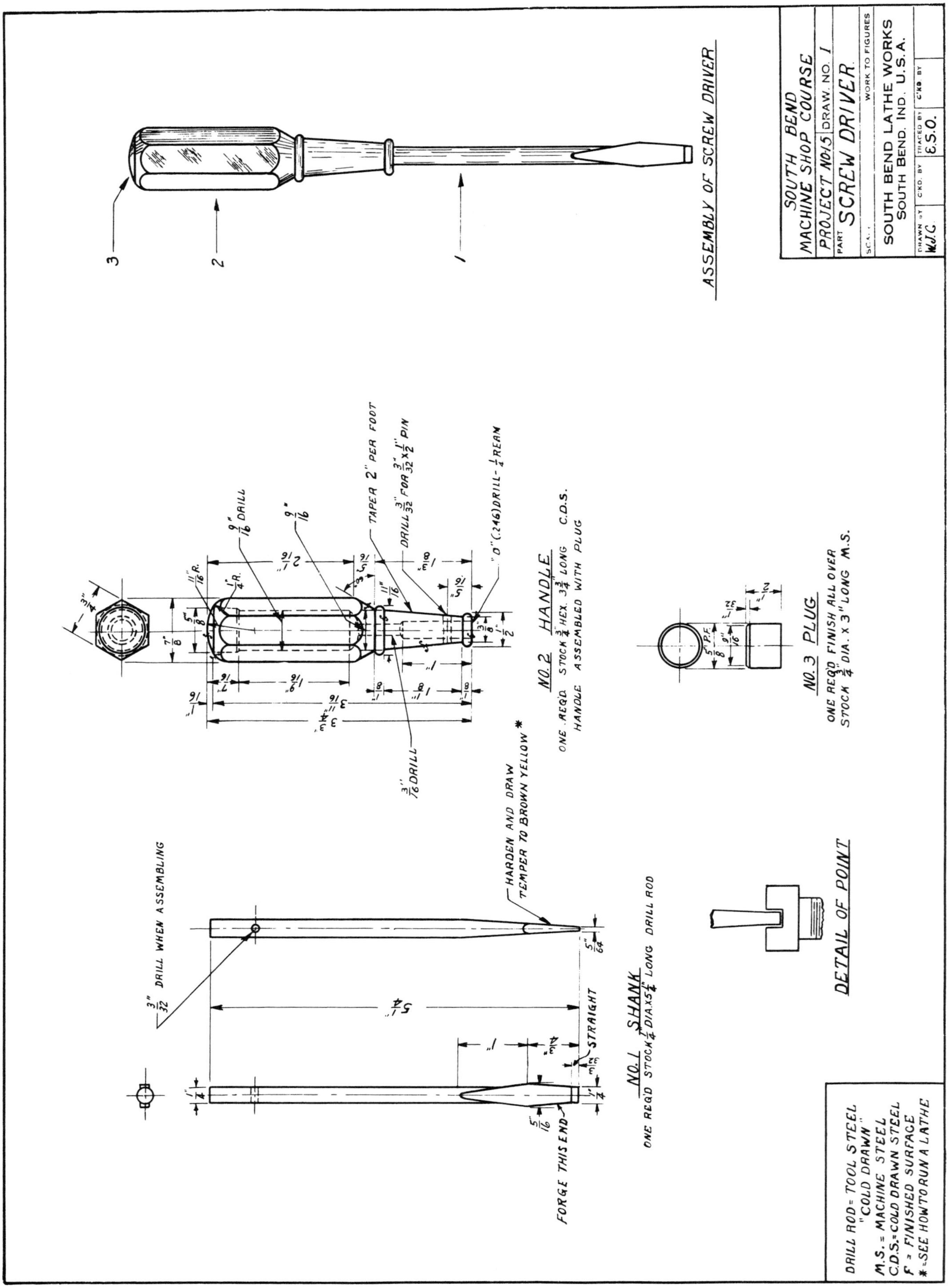
SOUTH BEND
MACHINE SHOP COURSE
PROJECT NO. 15
DRAW. NO. 1
PART SCREW DRIVER
WORK TO FIGURES
SOUTH BEND LATHE WORKS
SOUTH BEND, IND. U.S.A.
DRAWN BY W.J.C.
TRACED BY E.S.O.
ASSEMBLY OF SCREW DRIVER
9/16" DRILL
TAPER 2" PER FOOT
DRILL 3/32" FOR 3/32" X 1/2" PIN
"D" (.246) DRILL - 1/4 REAM
3/16" DRILL
NO. 2 HANDLE
ONE REQ'D STOCK 3/4" HEX. 3 3/4" LONG C.D.S.
HANDLE ASSEMBLED WITH PLUG
NO. 3 PLUG
ONE REQ'D FINISH ALL OVER
STOCK 3/4" DIA. X 3" LONG M.S.
DETAIL OF POINT
HARDEN AND DRAW
TEMPER TO BROWN YELLOW *
3/32" DRILL WHEN ASSEMBLING
STRAIGHT
FORGE THIS END
NO. 1 SHANK
ONE REQ'D STOCK 1/4" DIA. X 5 1/4" LONG DRILL ROD
DRILL ROD = TOOL STEEL "COLD DRAWN"
M.S. = MACHINE STEEL
C.D.S. = COLD DRAWN STEEL
F = FINISHED SURFACE
* = SEE HOW TO RUN A LATHE

SCREW DRIVER

MATERIAL REQUIRED:

Part No. 1—Drill rod ¼″ dia. x 5¼″ long. One required.
Part No. 2—Cold drawn steel ¾″ hex. x 3¾″ long. One required.
Part No. 3—Machinery steel ¾″ dia. x 3″ long. One required.
Part No. 4—Cold rolled steel 3/32″ dia. x 9/16″ long. One required.

OPERATIONS:

PART No. 1—SHANK

1. Select stock as per drawing.
2. Forge end to shape as per drawing.
3. File or grind to smooth finish.
4. File flat surfaces at point parallel for 3/32″ as per enlarged section of drawing.
5. Harden and draw temper for 1″ back from end as per drawing.*

NOTE: Do not drill pin hole in round end until instructed.

PART No. 2—HANDLE

1. Select stock as per drawing.
2. Place steel in 3-jaw chuck with 2 inches projecting and running true.*
3. Arrange belt for proper spindle speed.*
4. Set lathe tool for facing.*
5. Face end.*
6. Set small round nose tool for taper turning.*
7. Rough turn outside diameter of bead at large end of taper, using longitudinal hand carriage feed.*
8. Rough turn outside diameter of bead at small end. Use longitudinal hand carriage feed.
9. Set compound rest to secure the required taper, which is approximately 95 degrees with handle toward the tailstock.
10. Rough turn taper between beads, feeding toward the headstock with compound rest screw.*
11. Finish turn outside diameter of both beads.*
12. Finish turn tapered section using fine feed.
13. Chamfer corner between large diameter bead and hexagon section. Use a broad nose tool.
14. Round corners of both beads with a forming tool and file.
15. Center end of stock for drilling, with sharp pointed tool held in tool post.
16. Drill hole in end of stock as per drawing, with drill held in drill chuck in tailstock spindle.*
17. Insert machine reamer in drill chuck in place of drill.*
18. Ream hole in end for a light drive fit in Part No. 1—Shank, as per drawing.*
19. File machined surfaces lightly to remove tool marks.*
20. Polish machined surfaces with emery cloth and oil.*
21. Turn stock end for end in chuck, with 1 inch projecting and true as before.
22. Set lathe tool for facing.*
23. Face end to length as per drawing.*
24. Center end for drilling, with centering tool held in tool post.*
25. Drill 9/16″ hole in handle to depth as per drawing.
26. Continue through handle with 3/16″ drill to meet ¼″ hole in tapered end.
27. Set tool for boring.*
28. Bore end of handle for plug as per drawing, leaving .010″ to ream.
29. Ream as per drawing. Use chucking reamer held in tailstock spindle.*

PART No. 3—PLUG

1. Select stock as per drawing.
2. Place steel in 3-jaw chuck with 1 inch projecting and turning true.*
3. Arrange belt for proper spindle speed.*
4. Set lathe tool for facing.*
5. Face end.*
6. Set lathe tool for turning.*
7. Rough turn diameter.*
8. Finish turn diameter for driving fit into reamed hole in large end of Part No. 2—Handle.
9. Set tool for chamfering.
10. Chamfer end as per drawing.
11. Set tool for cutting off.
12. Cut plug off to length as per drawing, using oil on work and feeding tool slowly by hand.
13. Transfer work to bench. Drive plug into end of handle. (See footnote A.)
14. Place screw driver handle in lathe chuck with plug end extending 1 inch, and true as before.
15. Round end of handle and plug with forming tool as per drawing.
16. File lightly to remove tool marks and polish with fine emery cloth and oil.*
17. Transfer to bench. Drive shank into handle.
18. Centerpunch to drill hole through handle and shank for pin.
19. Transfer to drill press. Drill hole for pin as per drawing.
20. Countersink both ends of hole slightly.
21. Transfer to bench. Insert pin and rivet both ends.
22. File ends of pin flush to surface of handle.
23. Place back in lathe chuck and re-polish.

NOTE A: End of plug is chamfered for clearance at bottom of reamed hole. Oil plug and drive in chamfered end first.

*See book "How to Run a Lathe."

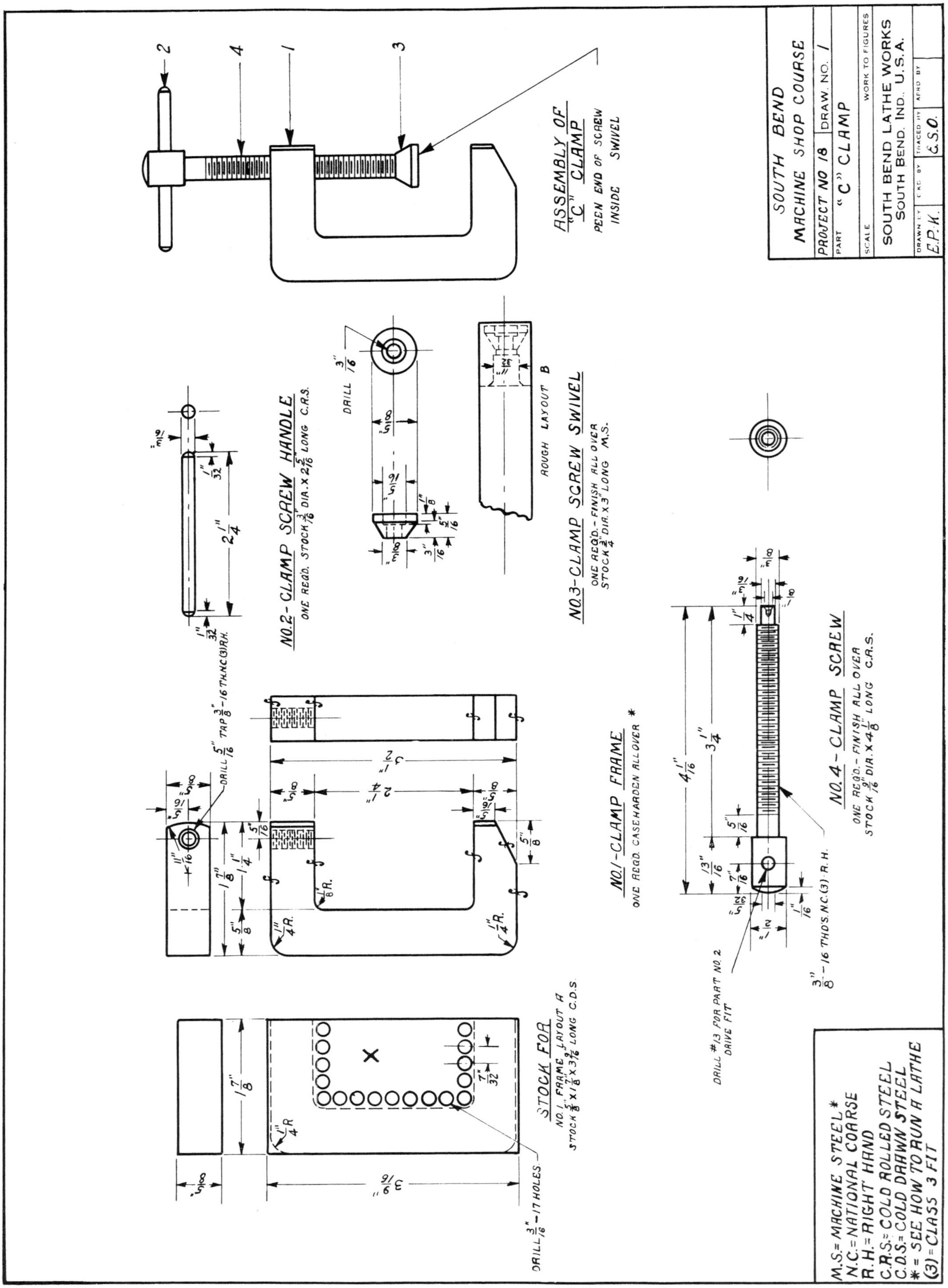
NO. 1 - CLAMP FRAME
ONE REQ'D. CASEHARDEN ALL OVER *
STOCK FOR
NO. 1 FRAME LAYOUT A
STOCK 5/8" X 1 7/8" X 3 9/16" LONG C.D.S.
DRILL 3/16" - 17 HOLES
DRILL 5/16" TAP 3/8" - 16 TH. N.C. (3) R.H.
NO. 2 - CLAMP SCREW HANDLE
ONE REQ'D. STOCK 3/16" DIA. X 2 5/16" LONG C.R.S.
NO. 3 - CLAMP SCREW SWIVEL
ONE REQ'D. - FINISH ALL OVER
STOCK 3/4" DIA. X 3" LONG M.S.
DRILL 3/16"
ROUGH LAYOUT B
ASSEMBLY OF "C" CLAMP
PEEN END OF SCREW INSIDE SWIVEL
NO. 4 - CLAMP SCREW
ONE REQ'D. - FINISH ALL OVER
STOCK 9/16" DIA. X 4 1/8" LONG C.R.S.
DRILL #13 FOR PART NO. 2 DRIVE FIT
3/8" - 16 THD'S. N.C. (3) R.H.
M.S. = MACHINE STEEL *
N.C. = NATIONAL COARSE
R.H. = RIGHT HAND
C.R.S. = COLD ROLLED STEEL
C.D.S. = COLD DRAWN STEEL
* = SEE HOW TO RUN A LATHE
(3) = CLASS 3 FIT
SOUTH BEND
MACHINE SHOP COURSE
PROJECT NO 18
DRAW. NO. 1
PART "C" CLAMP
SCALE
WORK TO FIGURES
SOUTH BEND LATHE WORKS
SOUTH BEND, IND., U.S.A.
DRAWN BY E.P.K.
CKD. BY
TRACED BY E.S.O.
APRD. BY

"C" CLAMP

MATERIAL REQUIRED:

Part No. 1—Cold drawn steel 5/8" x 1 7/8" x 3 9/16" long. One required.
Part No. 2—Cold rolled steel 3/16" dia. x 2 5/16" long. One required.
Part No. 3—Machinery steel 3/4" x 3" long. One required.
Part No. 4—Cold rolled steel 9/16" dia. x 4 1/8" long. One required.

OPERATIONS:

PART No. 1—FRAME

1. Select stock as per drawing. This cold drawn steel is stock size 5/8" x 1 7/8", and can be purchased on the market.
2. Square ends to length as per drawing layout A by planing or filing.
3. Lay off finish lines inside of blank as per drawing.
4. Lay off and centerpunch to drill for removing stock X. See drawing layout A.
5. Transfer to drill press. Drill to remove stock X as per drawing layout A. (See footnote A.)
6. Transfer to bench. Saw or drift and chip out light sections between holes to remove stock X.
7. Transfer to shaper. Place stock in vise with open side up. Set work square, and plane out inside of blank as per drawing. (See footnote B.)
8. Transfer to bench. Finish inside of blank with a file. Use a round file for fillets in corners.
9. Lay off and centerpunch to drill hole to be tapped for Part No. 4—Screw.
10. Lay off rounded outside corners and ends as per drawing.
11. Lay off to plane angle on anvil end of frame.
12. Transfer to drill press. Drill hole for Part No. 4—Screw as per drawing. (See footnote C.)
13. Transfer to grinding wheel. Rough grind rounded outside corners and ends.
14. Transfer to bench. Finish rounded outside corners and ends with a file.
15. Tap hole for Part No. 4—Clamp Screw. Start thread with a taper tap and finish with a plug tap.
16. Transfer to shaper, and plane angle on anvil end of frame.
17. Remove burrs and sharp corners with a fine file.
18. File surfaces smooth. Polish with emery cloth.*
19. Transfer to forge and caseharden all over.*

PART No. 2—HANDLE

1. Select stock as per drawing.
2. Place stock in drill chuck in headstock spindle.
3. Arrange belt for proper spindle speed.*
4. Face and round end with forming tool in one operation.
5. File and polish end with emery cloth.*
6. Turn stock end for end in chuck.
7. Face end to length as per drawing and round end.
8. File and polish end with emery cloth.*

PART No. 3—SWIVEL

1. Select stock as per drawing.
2. Place stock in 3-jaw universal chuck, projecting 1 inch and running true.*
3. Arrange belt for proper spindle speed.*
4. Set lathe tool for facing.*
5. Face end.*
6. Set small round nose tool for turning.*
7. Rough turn large diameter section to 21/32" in diameter.*
8. Set compound rest at an angle of 55 degrees to the right.
9. Set small round nose tool for taper turning.*
10. Turn off surplus stock back of taper as per rough layout B.
11. Rough turn angle on swivel, feeding with compound rest screw.*
12. Finish turn angle on swivel feeding as before.*
13. Finish turn large diameter section as per drawing. Use same tool and feed with longitudinal feed.
14. Center end of stock for drilling with sharp pointed tool held in tool post.
15. Drill into end of stock 1/2" deep, with 3/16" drill held in drill chuck in tailstock spindle.
16. Counterbore to depth as per drawing. Use 5/16" counter-bore with 3/16" pilot.
17. File machined surfaces lightly to remove tool marks. Polish with emery cloth.*
18. Set right hand tool and cut off swivel to length as per drawing.* Put oil on work. Feed tool slowly by hand.
19. Remove burr from ends of hole with a file, and hand scraper.

PART No. 4—CLAMP SCREW

1. Select stock as per drawing.
2. Lay off and center ends.*
3. Place stock on centers in lathe.* Drive with common lathe dog.* (See footnote D.)
4. Arrange belt for proper spindle speed.*
5. Set lathe tool for facing.*
6. Face end.*
7. Turn stock end for end on centers.
8. Face end to length as per drawing.
9. Set lathe tool for turning.*
10. Rough turn diameter of section to be threaded.*
11. Turn stock end for end on centers.
12. Rough turn diameter of head section.
13. Finish turn diameter of head section.*
14. Round end of head section with forming tool.
15. Turn stock end for end on centers. Place soft sheet brass around work to prevent marring of finished surface by lathe dog.*
16. Set tool for turning.*
17. Finish turn stock to 3/8" in diameter as per drawing.
18. Finish turn diameter of section to be threaded .010" under-size to insure a free fit on thread.
19. Set tool for facing.*
20. Face shoulder of head section.
21. Arrange gearing for thread cutting.*
22. Set tool for thread cutting.*
23. Cut thread to fit into tapped hole in Part No. 1—Frame.
24. Set tool for turning.*
25. Rough turn small diameter or pilot section.
26. Finish turn small diameter or pilot section for running fit in Part No. 3—Swivel, and to length as per drawing.
27. Set tool for facing.
28. Face shoulder of threaded section.
29. Transfer to bench. Lay off and centerpunch to drill hole through head section.
30. Transfer work to drill press.
31. Drill hole through section for driving fit for Part No. 2—Handle. (See footnote E.)

ORDER OF ASSEMBLY

1. Put oil on Part No. 2—Handle and drive same into hole in head of Part No. 4—Screw.
2. Screw Part No. 4 into Part No. 1—Frame.
3. Place Part No. 3—Swivel on pilot section of Part No. 4—Screw and peen end of screw. Swivel should turn freely after end of Part No. 4 is peened.

NOTE A: Drilling can be done in the lathe, using drill pad in tailstock spindle and drill held in drill chuck in head stock spindle.*

NOTE B: If shaper is not available, remove stock by grinding or chipping and filing.

NOTE C: Drilling can be done in lathe as follows: Lay off and centerpunch anvil end of frame in line with center of hole to be drilled for part No. 4—Screw, and drill between centers.*

NOTE D: A relieved center would simplify the facing operation.*

NOTE E: Drilling can be done in lathe, using crotch center in tailstock spindle, and drill held in drill chuck in head-stock spindle.*

*See book "How to Run a Lathe."

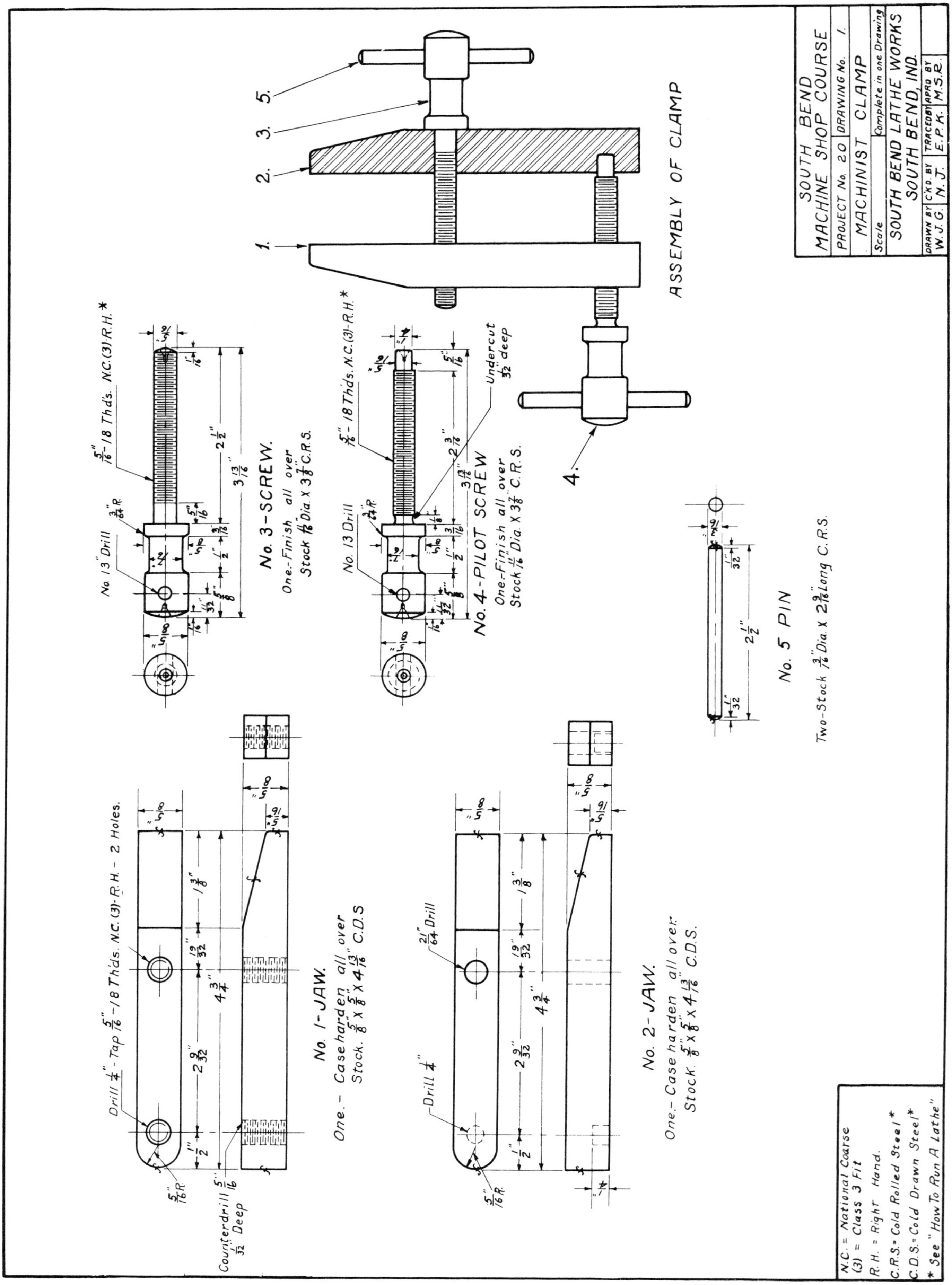
SOUTH BEND MACHINE SHOP COURSE
PROJECT No. 20
DRAWING No. 1
MACHINIST CLAMP
Scale
Complete in one Drawing
SOUTH BEND LATHE WORKS SOUTH BEND, IND.
DRAWN BY W.J.G.
CK'D. BY N.J.
TRACED BY E.P.K.
APP'D BY M.S.R.
ASSEMBLY OF CLAMP
No. 3-SCREW.
One-Finish all over Stock 11/16" Dia. X 3 7/8" C.R.S.
5/16"-18 Thd's. N.C.(3) R.H.*
No. 13 Drill
No. 4-PILOT SCREW
One-Finish all over Stock 11/16" Dia. X 3 7/8" C.R.S.
Undercut 1/32" deep
No. 5 PIN
Two-Stock 3/16" Dia. X 2 9/16" Long C.R.S.
No. 1-JAW.
One.- Caseharden all over Stock 5/8" X 5/8" X 4 13/16" C.D.S
Drill 1/4"-Tap 5/16"-18 Thd's. N.C.(3) R.H. - 2 Holes.
Counterdrill 1/32" Deep
No. 2-JAW.
One.- Caseharden all over Stock 5/8" X 5/8" X 4 13/16" C.D.S.
21/64" Drill
Drill 1/4"
N.C. = National Coarse
(3) = Class 3 Fit
R.H. = Right Hand.
C.R.S. = Cold Rolled Steel*
C.D.S. = Cold Drawn Steel*
* See "How To Run A Lathe"

MACHINIST'S CLAMP

MATERIAL REQUIRED:

Part Nos. 1 and 2—Cold drawn steel 5/8" x 5/8" x 4 13/16" long. Two required.
Part Nos. 3 and 4—Cold rolled steel 11/16" dia. x 3 7/8" long. Two required.
Part No. 5—Cold rolled steel 3/16" dia. x 2 9/16" long. Two required.

OPERATIONS:

PARTS Nos. 1 and 2—JAWS

1. Select two pieces of stock as per drawing. Perform following operations on both parts except as directed otherwise.
2. Place stock in 4-jaw chuck with 1 inch projecting. Adjust jaws so stock runs true.*
3. Arrange belt for proper spindle speeds.*
4. Set lathe tool for facing.*
5. Face one end only.*
6. Transfer work to bench.
7. Lay off bevel at faced end of stock.
8. Transfer to shaper and plane beveled section. (See footnote A.)
9. Transfer to bench. Lay off and file rounded ends to shape as per drawing.
10. Lay off and centerpunch to drill for screw holes in jaw No. 1.
11. Place jaw No. 1 on top of jaw No. 2 in position they will occupy when assembled and clamp them together.
12. Transfer to drill press. Drill hole for tap at beveled end, drilling through both pieces.* (See footnote B.)
13. Drill hole for tap at opposite end through jaw No. 1 and part way into jaw No. 2 as per drawing.
14. Redrill hole through jaw No. 2 at beveled end as per drawing.
15. Square bottom of blind hole in jaw No. 2 using a drill ground flat at point.
16. Transfer to bench vise. Tap holes through jaw No. 1.
17. Transfer to drill press. Countersink tapped hole in jaw No. 1 with a 5/16" drill to depth of ½ thread.
18. Transfer to bench vise. Break sharp corners with a file. File and polish all surfaces.
19. Transfer to forge. Caseharden all over.*

PARTS Nos. 3 and 4—SCREWS

1. Select two pieces of stock as per drawing. Perform following operations on both parts except as directed otherwise.
2. Lay off and center ends.*
3. Place on centers in lathe.* Drive with common lathe dog.* (See footnote C.)
4. Arrange belt for proper spindle speed.*
5. Set lathe tool for facing.*
6. Face end.*
7. Turn stock end for end on centers.
8. Face end to length as per drawing.
9. Set lathe tool for turning.*
10. Rough turn section to be threaded.*
11. Turn work end for end on centers.
12. Rough turn large diameter section to rough size for shoulder.
13. Rough turn recess section, using round nose tool.
14. Finish turn large diameter section.*
15. Finish turn recessed section, using small round nose tool ground to form fillets at shoulders.*
16. Set form facing tool and form radius on head end.
17. File radius on collar as per drawing. File machined surfaces lightly to remove tool marks and break sharp corners.*
18. Polish finished section with emery cloth and oil.
19. Turn work end for end on centers. Place soft sheet brass around finished work to prevent marring by lathe dog.*
20. Set tool for turning.*
21. Finish turn section to be threaded. Turn .010" undersize to insure a free fit.
22. Cut threading recess on No. 4—Screw as per drawing with narrow round nose tool. Face shoulder as per drawing.
23. Arrange gears for cutting thread as per drawing.*
24. Set tool for thread cutting.*
25. Cut thread to fit tapped holes in jaw No. 1.*
26. Round end of Screw No. 3 with forming tool.
27. Turn small section at end of thread on screw No. 4 for pilot. This should be a running fit in blind hole in jaw No. 2.
28. Transfer to bench. Centerpunch to drill hole through head section.
29. Transfer to drill press. Drill hole through head section for driving fit for Part No. 5—Pin. (See footnote D.)

PART No. 5—PIN

1. Select stock as per drawing.
2. Place stock in drill chuck in headstock spindle with 1 inch projecting.
3. Round end with forming tool as per drawing.
4. Turn stock end for end projecting as before.
5. Round end as before.
6. Transfer to bench. Assemble clamp as per drawing.

NOTE A: If shaper is not available stock may be removed by drilling, or grinding and filing.

NOTE B: Drilling can be done in the lathe using drill pad in tailstock spindle, and drill held in drill chuck in headstock spindle.*

NOTE C: A relieved center would simplify the facing operation.*

NOTE D: Drilling may be done in lathe using crotch center in tailstock spindle and drill held in drill chuck in headstock spindle.*

*See book "How to Run a Lathe."

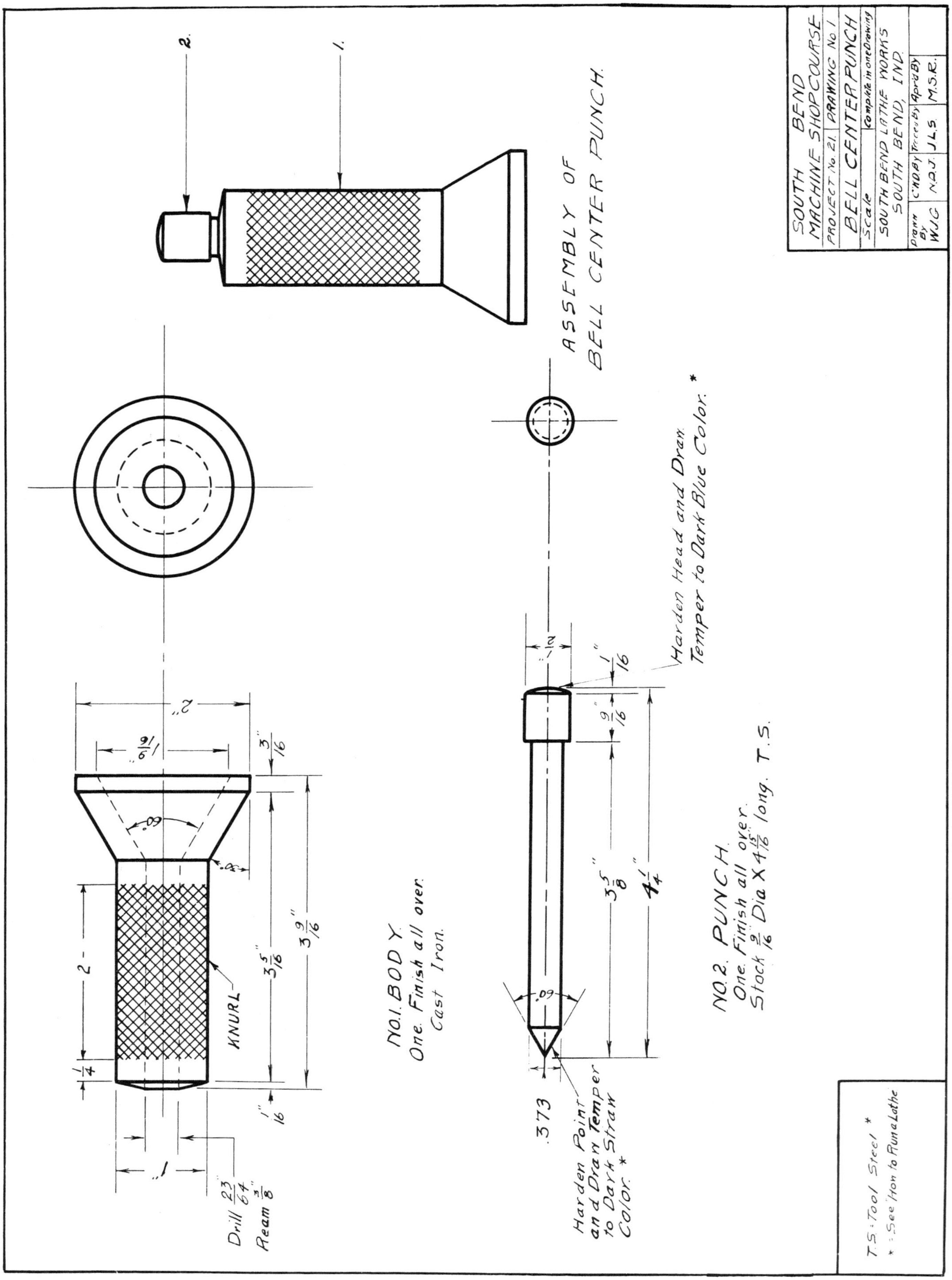
SOUTH BEND MACHINE SHOP COURSE
PROJECT No. 21. DRAWING No. 1
BELL CENTER PUNCH
Scale
Complete in one Drawing
SOUTH BEND LATHE WORKS
SOUTH BEND, IND.
Drawn By W.J.G.
Ch'kd By N.D.J.
Traced By J.L.S.
App'd By M.S.R.
ASSEMBLY OF BELL CENTER PUNCH
Harden Head and Draw Temper to Dark Blue Color. *
NO. 1 BODY. One. Finish all over. Cast Iron.
KNURL
Drill 23/64" Ream 3/8"
NO. 2 PUNCH. One. Finish all over. Stock 9/16" Dia X 4 15/16" long. T.S.
Harden Point and Draw Temper to Dark Straw Color. *
.373
T.S.: Tool Steel *
*: See "How to Run a Lathe"

BELL CENTER PUNCH

MATERIAL REQUIRED:

Part No. 1—Cast iron. One required.
Part No. 2—Tool steel 9/16″ dia. x 4 15/16″ long. One required.

OPERATIONS:

PART No. 1—BODY

1. Place casting in 4-jaw chuck with bell end projecting 1¾″. Adjust jaws so casting runs true.*
2. Arrange belt for proper spindle speed.*
3. Set lathe tool for facing.*
4. Face end.*
5. Set lathe tool for turning.*
6. Rough turn outside diameter of bell.
7. Set compound rest at proper angle for turning taper on outside of bell, which is 60 degrees to the right.
8. Set lathe tool for taper turning.*
9. Rough turn taper on outside of bell, feeding with compound rest screw.
10. Set tool for taper boring.
11. With compound rest setting as before, rough bore taper inside of bell.
12. Face center section inside of bell to depth as per drawing.
13. Finish bore taper inside of bell.
14. Center for drilling with sharp pointed tool held in tool post.*
15. Drill hole through center of casting as per drawing, holding drill in drill chuck in tail-stock spindle.*
16. Insert rose chucking reamer in drill chuck in place of drill.*
17. Ream hole through center of casting, as per drawing.*
18. Remove casting from chuck and transfer to arbor press.
19. Remove all dirt and oil inside of hole.
20. Oil lathe mandrel, and press same in hole in casting.
21. Place on centers in lathe, driving with common lathe dog on mandrel at bell end.*
22. Set lathe tool for facing.*
23. Face end to length as per drawing.
24. Set lathe tool for turning.*
25. Rough turn small diameter section.
26. Finish turn small diameter section.
27. Chamfer end as per drawing.
28. File lightly to remove tool marks and polish with fine emery cloth.*
29. Turn work end for end on centers.
30. Set lathe tool for taper turning.*
31. Finish turn outside diameter of bell.
32. With compound rest still set at an angle of 60 degrees to the right, finish turn taper on inside of bell.
33. File lighly to remove tool marks and polish with fine emery cloth.*
34. Transfer to arbor press and remove mandrel.

PART No. 2—PUNCH

1. Select stock as per drawing.
2. Lay off and center ends.*
3. Place on centers in lathe. Drive with common lathe dog.* (See footnote A.)
4. Arrange belt for proper spindle speed.*
5. Set lathe tool for facing.*
6. Face end.*
7. Turn stock end for end on centers.
8. Face end to length as per drawing, plus 5/8″ for removing the centers.
9. Set lathe tool for turning.*
10. Rough turn small diameter section to length as per drawing, plus 5/16″ for removing the center.*
11. Rough turn large diameter section.
12. Finish turn large diameter section.
13. File lightly to remove tool marks and polish with fine emery cloth.*
14. Turn stock end for end on centers. Place a piece of soft brass under the dog screw to prevent marring the finished surface.*
15. Finish turn small diameter section as per drawing.
16. Set lathe tool for facing.*
17. Face shoulder of large diameter section.
18. File small diameter section for a slip fit for reamed hole in Part No. 1—Body, and polish with fine emery cloth.*
19. Remove work from centers and place in lathe chuck with large end projecting. Adjust jaws so work runs true.*
20. Face end to length as per drawing, removing center hole.
21. Round end with forming tool.
22. Turn work end for end in chuck with one inch projecting and true as before.
23. Set lathe tool for facing.*
24. Face end to length as per drawing, removing center hole.
25. Set compound rest at proper angle for turning point, which is 120 degrees to the right.
26. Set lathe tool for taper turning.*
27. Turn angle of point as per drawing.
28. Transfer to forge. Harden head and draw temper as per drawing.
29. Harden point and draw temper as per drawing.*
30. Transfer to bench and assemble body and punch.

NOTE A: A relieved center would simplify the facing operation.*

*See book "How to Run a Lathe."

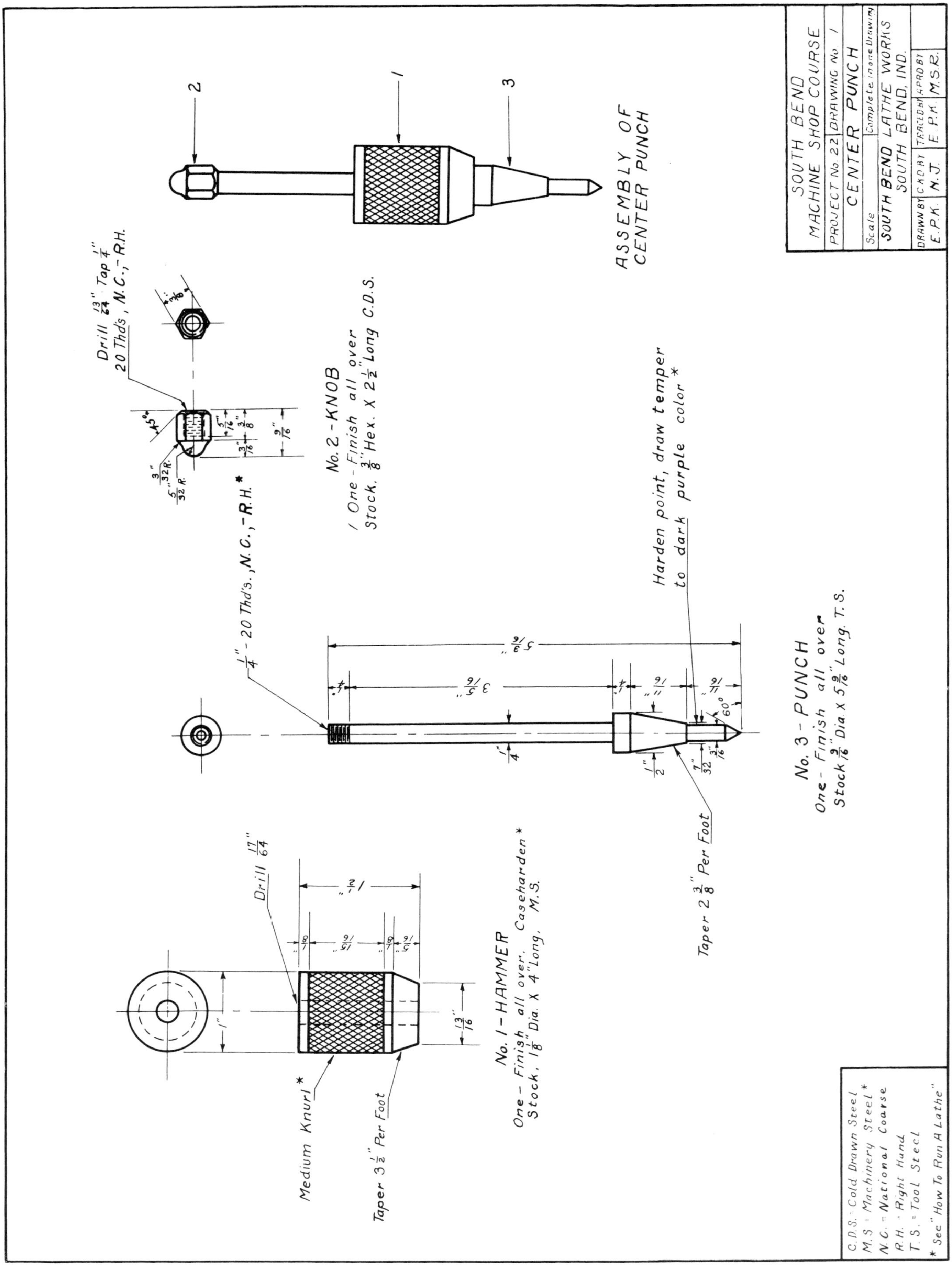
SOUTH BEND MACHINE SHOP COURSE
PROJECT No. 22
DRAWING No. 1
CENTER PUNCH
Scale
Complete in one Drawing
SOUTH BEND LATHE WORKS
SOUTH BEND, IND.
DRAWN BY E.P.K.
CKD BY N.J.
TRACED BY E.P.K.
APPRO BY M.S.R.
ASSEMBLY OF CENTER PUNCH
Drill 13/64" Tap 1/4"
20 Thd's, N.C.,-R.H.
No. 2 - KNOB
1 One - Finish all over
Stock. 3/8" Hex. X 2 1/2" Long C.D.S.
1/4" - 20 Thd's., N.C., -R.H.*
Harden point, draw temper
to dark purple color*
No. 3 - PUNCH
One - Finish all over
Stock 9/16" Dia. X 5 9/16" Long, T.S.
Taper 2 3/8" Per Foot
Drill 17/64"
Medium Knurl*
Taper 3 1/2" Per Foot
No. 1 - HAMMER
One - Finish all over, Caseharden*
Stock, 1 1/8" Dia. X 4" Long, M.S.
C.D.S. = Cold Drawn Steel
M.S. = Machinery Steel*
N.C. = National Coarse
R.H. = Right Hand
T.S. = Tool Steel
* See "How To Run A Lathe"

CENTER PUNCH WITH SLIDING SLEEVE HAMMER

MATERIAL REQUIRED:

Part No. 1—Machinery steel $1\frac{1}{8}$" dia. x 4" long. One required.
Part No. 2—Cold drawn steel $\frac{3}{8}$" hex. x $2\frac{1}{2}$" long. One required.
Part No. 3—Tool steel $\frac{9}{16}$" dia. x $5\frac{9}{16}$" long. One required.

OPERATIONS:

PART No. 1—HAMMER

1. Select stock as per drawing.
2. Place stock in 4-jaw independent chuck projecting two inches. Adjust jaws so stock runs true.*
3. Arrange belt for proper spindle speed.*
4. Set lathe tool for facing.*
5. Face end.*
6. Center drill end of stock with center drill held in drill chuck in tailstock spindle.*
7. Remove drill chuck and place hardened center in tailstock spindle.
8. Adjust tailstock to support end of work.*
9. Set lathe tool for turning.*
10. Rough turn diameter.*
11. Finish turn diameter.*
12. Set compound rest at proper angle to obtain a taper of $3\frac{1}{2}$" per foot. This is approximately 82 degrees to the left.
13. Turn taper on nose of work. Feed tool with compound rest screw.*
14. File machined surfaces lightly to remove tool marks. Round corners with file. Polish with emery cloth.*
15. Set tool for knurling.*
16. Knurl large diameter section as per drawing.*
17. Drill into end of stock with drill as per drawing. Hold drill in drill chuck in tailstock spindle.*
18. Set right-hand tool for cutting off.*
19. Cut off work to length as per drawing, plus $\frac{1}{32}$" to finish end. Put oil on work and feed tool slowly by hand supporting work with tailstock center.*
20. Place work in lathe chuck with unfinished end projecting $\frac{1}{4}$" and true as before. Protect knurl with soft sheet brass.
21. Set lathe tool for facing.*
22. Face end to length as per drawing.*
23. Round corner with a file. Polish face with fine emery cloth.
24. Transfer to forge and caseharden all over.*

PART No. 2—KNOB

1. Select stock as per drawing.
2. Place stock in 3-jaw universal chuck, with one inch projecting and running true.*
3. Arrange belt for proper spindle speed.*
4. Set lathe tool for facing.*
5. Face end.*
6. Set tool for chamfering.
7. Chamfer corners.
8. Center drill end of stock. Hold center drill in drill chuck in tailstock spindle.*
9. Drill to size and depth as per drawing.
10. Tap as per drawing. Hold tap in drill chuck in tailstock spindle. Pull belt by hand.
11. Set right-hand tool for cutting off.*
12. Cut off work to length as per drawing, plus $\frac{1}{32}$" to finish end. Put oil on work and feed tool slowly by hand.
13. Place work in lathe chuck with unfinished end out and running true as before.
14. Set tool for facing.*
15. Face end to length as per drawing.*
16. Round end and form fillet with forming tools as per drawing.
17. File lightly to remove tool marks.*
18. Polish with emery cloth.*

PART No. 3—PUNCH

1. Select stock as per drawing.
2. Lay off and center ends.*
3. Place stock on centers in lathe. Drive with common lathe dog.* (See footnote A.)
4. Arrange belt for proper spindle speed.*
5. Set lathe tool for facing.*
6. Face end.*
7. Turn stock end for end on centers.
8. Face end to length as per drawing, plus $\frac{5}{16}$" for removing the center hole in punch end.
9. Set lathe tool for turning.*
10. Rough turn $\frac{1}{4}$" diameter section as per drawing.*
11. Rough turn large diameter section.
12. Finish turn large diameter section.*
13. Finish turn $\frac{1}{4}$" diameter section as per drawing. Allow .003" for filing to size.
14. Finish turn short section to be threaded .010" undersize to insure free fit on thread.
15. Set lathe tool for facing.*
16. Face shoulder of large diameter to secure exact length on the $\frac{1}{4}$" diameter section.
17. File finished surfaces to remove tool marks and secure exact size as per drawing. Polish with emery cloth.*
18. Arrange gearing for thread cutting.*
19. Set tool for thread cutting.*
20. Cut thread to fit into tapped hole in Part No. 2—Knob.*
21. Remove work from centers and place same in lathe chuck. Grip work on $\frac{1}{4}$" diameter with punch end projecting. Adjust so finished large diameter runs true.*
22. Set lathe tool for turning.*
23. Rough turn $\frac{3}{16}$" dia. on punch end.*
24. Finish turn $\frac{3}{16}$" dia. on punch end.*
25. Set lathe tool for facing.*
26. Face shoulder of large diameter section to length as per drawing.*
27. Face punch end to length as per drawing. This will remove the center hole.
28. Set compound rest at the proper angle to obtain a taper of $2\frac{3}{8}$" per foot. This is approximately 96 degrees to the right.
29. Set lathe tool for taper turning.*
30. Turn tapered section as per drawing, feeding with compound rest screw.*
31. Set compound rest at an angle of 120 degrees to the right.
32. Turn angle of point as per drawing.
33. File machined surfaces lightly to remove tool marks. Polish with fine emery cloth.*
34. Transfer to forge. Harden point for one inch back from end, and draw temper as per drawing.

NOTE A: A relieved center would simplify the facing operation.*

*See book "How to Run a Lathe."

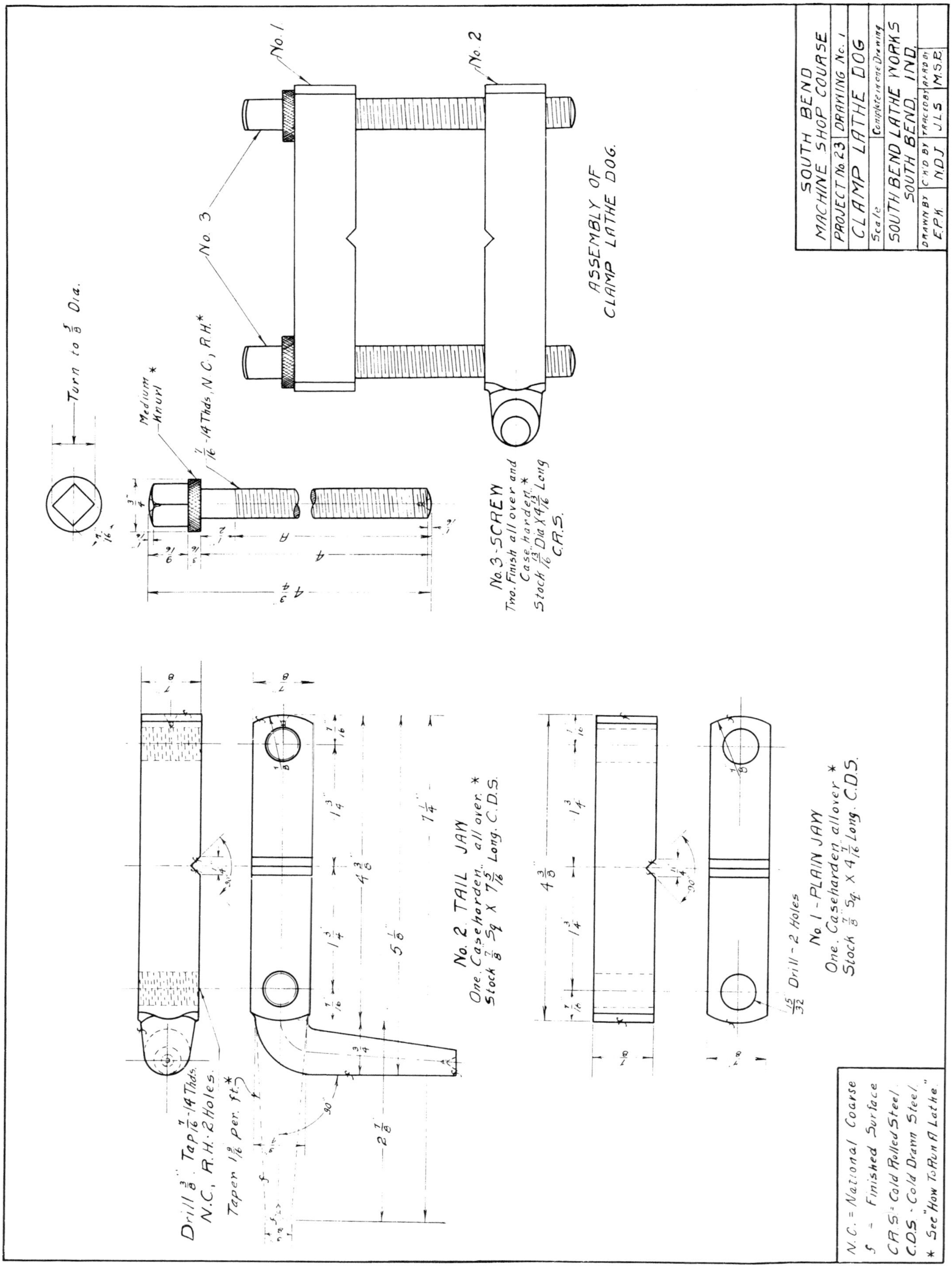
SOUTH BEND
MACHINE SHOP COURSE
PROJECT No 23
DRAWING No. 1
CLAMP LATHE DOG
Scale
Complete in one Drawing
SOUTH BEND LATHE WORKS
SOUTH BEND, IND.
DRAWN BY E.P.K.
CH'KD BY N.D.J.
TRACED BY J.L.S.
APP'D BY M.S.R.
ASSEMBLY OF CLAMP LATHE DOG.
No. 1
No. 2
No. 3
Turn to 5/8 Dia.
Medium Knurl *
7/16-14 Thds, N.C, R.H.*
No. 3-SCREW
Two. Finish all over and Case harden.*
Stock 13/16 Dia X 4 13/16 Long
C.R.S.
Drill 3/8" Tap 7/16-14 Thds. N.C, R.H. 2 Holes.
Taper 1 9/16 per ft.*
No. 2 TAIL JAW
One. Caseharden all over.*
Stock 7/8 Sq X 7 5/16 Long. C.D.S.
15/32 Drill - 2 Holes
No. 1 - PLAIN JAW
One. Caseharden all over *
Stock 7/8 Sq. X 4 7/16 Long. C.D.S.
N.C. = National Coarse
f = Finished Surface
C.R.S. = Cold Rolled Steel
C.D.S. - Cold Drawn Steel
* See "How To Run A Lathe."

CLAMP LATHE DOG

MATERIAL REQUIRED:

Part No. 1—Cold drawn steel 7/8" square x 4 7/16" long. One required.
Part No. 2—Cold drawn steel 7/8" square x 7 5/16" long. One required.
Part No. 3—Cold rolled steel 13/16" dia. x 4 13/16" long. Two required.

OPERATIONS:

PART No. 1—PLAIN JAW

1. Select stock as per drawing.
2. Lay off and mark location of 90 degree notch in center of stock.
3. Locate and draw a line through the center lengthwise of the stock to intersect the lines already drawn for location of 90 degree notch.
4. Center punch lightly at intersection of the two center lines.
5. Starting from this center, lay off and mark location of holes to be drilled. Lay off to round ends of stock.
6. Transfer to milling machine or shaper and machine the 90 degree notch. (See footnote A.)
7. Transfer to grinding wheel. Rough grind rounded ends to proper shape.
8. Transfer to drill press. Drill holes as per drawing. (See footnote B.)
9. Transfer to bench. Finish rounded ends with a file.
10. Remove all burrs and sharp corners with a file.
11. File all surfaces smooth. Polish with fine emery cloth.*
12. Transfer to forge and caseharden all over.*

PART No. 2—TAIL JAW

1. Select stock as per drawing.
2. Lay off and center ends.*
3. Place stock on centers in lathe.* Drive with common lathe dog.* (See footnote C.)
4. Arrange belt for proper spindle speed.*
5. Set lathe tool for facing.*
6. Face end.*
7. Turn stock end for end on centers.
8. Face end to length as per drawing.
9. Arrange lathe for taper turning.*
10. Set lathe tool for taper turning.*
11. Turn off surplus stock.*
12. Take light trial cut over tapered section.
13. Caliper both ends of tapered section to see if taper is correct. If taper is not correct adjust set-over to correct error.
14. Rough turn tapered section.
15. Finish turn tapered section as per drawing.
16. File tapered section lightly to remove tool marks.*
17. Polish with emery cloth.*
18. Transfer to bench. Lay off for center of notch, and to round end, as on jaw No. 1.
19. Transfer to milling machine or shaper and machine notch. (See footnote A.)
20. Grind and file rounded end as on jaw No. 1.
21. Clamp jaw No. 1 on top of jaw No. 2 in position they will occupy when assembled. See that the 90 degree notches are opposite each other.
22. Transfer to drill press. Using plain jaw as a template, spot centers of holes in tail jaw with 15/32" drill. (See footnote B.)
23. Drill holes through tail jaw as per drawing.
24. Transfer to bench with parts still clamped together. Tap holes in tail jaw, using holes in plain jaw to guide tap.
25. Remove burrs and sharp corners with a file.
26. File all surfaces smooth. Polish with emery cloth.*
27. Transfer to forge. Heat and bend tail as per drawing. Caseharden all over.*

PART No. 3—SCREW

1. Select two pieces of stock as per drawing. Perform following operations on both parts.
2. Lay off and center ends.*
3. Place stock on centers in lathe. Drive with common lathe dog.* (See footnote C.)
4. Arrange belt for proper spindle speed.*
5. Set lathe tool for facing.*
6. Face end.*
7. Turn stock end for end on centers.
8. Face end to length as per drawing.
9. Set lathe tool for turning.*
10. Rough turn small diameter section.*
11. Turn stock end for end on centers.
12. Rough turn large diameter section.
13. Rough turn head section.
14. Finish turn large diameter section.
15. Finish turn head section.
16. Set tool for knurling.*
17. Knurl large diameter section as per drawing.*
18. Round end of head section with forming tool.
19. Turn stock end for end on centers.
20. Set tool for turning.*
21. Finish turn small diameter section.
22. Turn section A of small diameter section .010" undersize to insure free fit on thread.
23. Face shoulder of knurled section as per drawing.
24. Arrange gearing for thread cutting.*
25. Set tool for thread cutting.*
26. Cut thread to fit tapped holes in Part No. 2—Tail Jaw.*
27. Round end of threaded section with forming tool.
28. Transfer to milling machine.
29. Straddle mill square on head section as per drawing. (See footnote D.)

NOTE A: If milling machine or shaper is not available, remove stock to form notch with a square file.

NOTE B: Drilling can be done in the lathe, using drill pad in tailstock spindle and drill held in drill chuck in headstock spindle.*

NOTE C: A relieved center would simplify the facing operation.*

NOTE D: If milling machine is not available lay off and square ends by grinding and filing.

*See book "How to Run a Lathe."

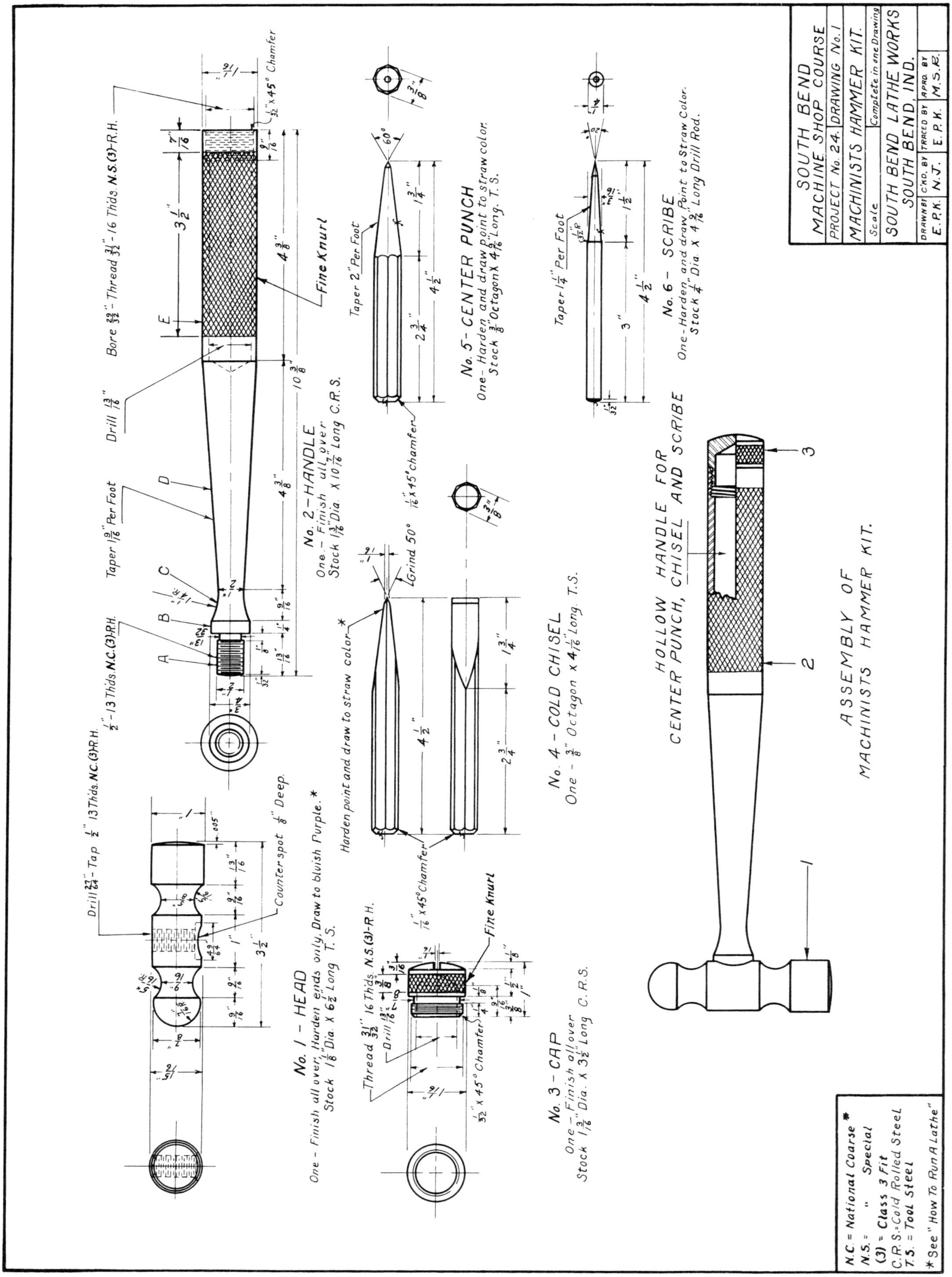
SOUTH BEND
MACHINE SHOP COURSE
PROJECT No. 24. DRAWING No. 1
MACHINISTS HAMMER KIT.
Scale
Complete in one Drawing
SOUTH BEND LATHE WORKS
SOUTH BEND, IND.
DRAWN BY E.P.K. CH'KD. BY N.J. TRACED BY E.P.K. APPD. BY M.S.R.
No. 1 - HEAD
One - Finish all over, Harden ends only, Draw to bluish Purple. *
Stock 1 1/8" Dia. X 6 1/2" Long T. S.
No. 2 - HANDLE
One - Finish all over
Stock 1 3/16" Dia. X 10 7/16" Long C. R. S.
Fine Knurl
Bore 29/32 - Thread 3 1/32" - 16 Thds. N.S.(3)-R.H.
1/32" X 45° Chamfer
Drill 13/16"
Taper 1 9/16" Per Foot
1/2" - 13 Thds. N.C.(3)-R.H.
Drill 27/64 - Tap 1/2" 13 Thds. N.C.(3)-R.H.
Counter spot 1/8" Deep.
No. 3 - CAP
One - Finish all over
Stock 1 3/16" Dia. X 3 1/2" Long C. R. S.
Thread 31/32" 16 Thds. N.S.(3)-R.H.
Drill 13/16
1/32" X 45° Chamfer
No. 4 - COLD CHISEL
One - 3/8" Octagon X 4 1/16" Long. T.S.
Harden point and draw to straw color. *
Grind 50°
1/16" X 45° Chamfer
No. 5 - CENTER PUNCH
One - Harden and draw point to straw color.
Stock 3/8" Octagon X 4 9/16" Long. T. S.
Taper 2" Per Foot
1/16" X 45° chamfer
No. 6 - SCRIBE
One - Harden and draw Point to Straw Color.
Stock 1/4" Dia. X 4 9/16" Long Drill Rod.
Taper 1 1/4" Per Foot
HOLLOW HANDLE FOR CENTER PUNCH, CHISEL AND SCRIBE
ASSEMBLY OF MACHINISTS HAMMER KIT.
N.C. = National Coarse *
N.S. = " Special
(3) = Class 3 Fit
C.R.S. = Cold Rolled Steel
T.S. = Tool Steel
* See "How To Run A Lathe"

MACHINIST'S HAMMER KIT

MATERIAL REQUIRED:

Part No. 1—Head. Tool steel 1⅛″ dia. x 6½″ long. One required.
Part No. 2—Handle. Cold rolled steel 1³⁄₁₆″ dia. x 10⁷⁄₁₆″ long. One required.
Part No. 3—Cap. Cold rolled steel 1³⁄₁₆″ dia. x 3½″ long. One required.
Part No. 4—Cold Chisel. Tool steel ⅜″ octagon x 4 ¹⁄₁₆″ long. One required.
Part No. 5—Center Punch. Tool steel ⅜″ octagon x 4⁹⁄₁₆″ long. One required.
Part No. 6—Scribe. Drill rod ¼″ dia. x 4⁹⁄₁₆″ long. One required.

OPERATIONS:

PART No. 1—HEAD

1. Select stock as per drawing.
2. Place stock in 4-jaw independent chuck, with 4½″ projecting. Adjust jaws so stock runs true.*
3. Arrange belt for proper spindle speed.*
4. Set lathe tool for facing.*
5. Face end.*
6. Center drill end of stock with center drill held in drill chuck in tailstock spindle.*
7. Remove drill chuck and place hardened center in tailstock spindle.*
8. Adjust tailstock to support end of work.*
   NOTE: The ball end of this part is to be formed on end of stock supported by tailstock. Allow ⁵⁄₁₆″ stock on ball end for removing the center when the ball is formed.
9. Set lathe tool for turning.*
10. Rough turn large diameter section.
11. Rough turn diameter of center section.
12. Rough turn small diameter section.
13. Finish turn the three diameters as per drawing. File lightly to remove tool marks. Polish with emery cloth.*
14. Lay off and mark location of the concave grooves as per drawing.
15. Rough turn both grooves, using a round nose tool.
16. Finish turn both grooves as per drawing, using a forming tool. Polish grooves with emery cloth.
17. Place center rest in position with jaws in line with ¹⁵⁄₁₆″ diameter section of work and clamp center rest to lathe bed.*
18. Adjust center rest jaws to center up on ¹⁵⁄₁₆″ diameter section and withdraw tailstock. Put oil on work to lubricate jaws.
19. Set round nose tool for facing.
20. Face end to length as per drawing. This operation should remove the center hole from end of work.
21. Rough turn ball with same tool.
22. Finish turn ball with forming tools. File ball lightly and polish with emery cloth.*
23. Remove center rest from lathe.
24. Set right hand tool for cutting off.*
25. Cut off work to length as per drawing, plus ¹⁄₃₂″ for finishing the end.
26. Place work in lathe chuck. Use ¹⁄₃₂″ sheet brass around work to prevent marring of surface by chuck jaws. Grip on center section, allowing large end to project 1″, and true as before.
27. Set lathe tool for facing.
28. Face end to length as per drawing.
29. Round face as per drawing.
30. File face to remove tool marks. Polish with emery cloth.*
31. Transfer to bench. Lay off and centerpunch to drill, tap and counterbore for Part No. 2—Handle.
32. Transfer to drill press. Drill and counterbore as per drawing. (See footnote A.)
33. Transfer back to bench. Tap hole as per drawing. Start thread with taper tap, and finish with a plug tap.
34. Transfer to forge. Harden ends only. Draw temper as per drawing.*

PART No. 2—HANDLE

1. Select stock as per drawing.
2. Lay off and center ends.*
3. Place stock on centers in lathe. Drive with common lathe dog.* (See footnote B.)
4. Arrange belt for proper spindle speed.*
5. Set lathe tool for facing.*
6. Face end.*
7. Turn stock end for end on centers.
8. Face end to length as per drawing.
9. Set lathe tool for turning.
10. Rough turn section E.*
11. Turn work end for end on centers.
12. Rough turn section B.
13. Rough turn section A.

*See book "How to Run a Lathe."

MACHINIST'S HAMMER KIT (Continued)

OPERATIONS:

PART No. 2—HANDLE (Continued)

14. Arrange lathe to obtain a taper of 1⁹⁄₁₆" per foot.*
15. Set lathe tool for taper turning.*
16. Take light trial cut over tapered section D.
17. Caliper at both ends of section D to test accuracy of taper. If taper is not correct adjust setting to correct error.
18. Rough turn tapered section D.
19. Rough turn fillet C at small end of tapered section D.
20. Finish turn tapered section D, as per drawing.*
21. Finish turn fillet C with a forming tool.
22. Arrange lathe for straight turning.*
23. Turn work end for end on centers.
24. Finish turn and knurl section E.
25. Turn work end for end on centers. Place soft sheet brass around finished end to prevent marring of surface by lathe dog.*
26. Finish turn section B.
27. Finish turn section A .010" undersize to insure a perfect fit on thread.
28. Set lathe tool for facing.*
29. Face shoulder of section B.
30. Set tool for recessing.*
31. Cut threading recess between sections A & B.
32. Arrange gearing for cutting thread as per drawing.*
33. Set tool for thread cutting.*
34. Cut thread to fit into tapped hole in Part No. 1—Head. Round end of threaded section with forming tool, as per drawing, not touching center hole.
35. Turn work end for end on centers. Drive with dog attached to ½" nut screwed onto threaded end of handle.*
36. Place center rest into position with jaws ½" from end of section E.*
37. Adjust jaws to center up on diameter of section E. Put oil on work to lubricate jaws.
38. Arrange belt lacing around lathe dog and through slots of face plate, for holding work against the live center.*
39. Withdraw tailstock. Remove hardened center and insert drill chuck in tailstock spindle.*
40. Center end of handle for drilling, with centering tool held in tool post.*
41. Drill end of handle as per drawing, holding drill in drill chuck in tailstock spindle.*
42. Set small tool for boring.*
43. Bore end of handle for thread as per drawing.
44. Arrange gearing for cutting internal thread as per drawing.*
45. Set tool for internal thread cutting.*
46. Cut thread to size as per drawing.*
47. Set tool for chamfering.
48. Chamfer end of thread as per drawing.

PART No. 3—CAP

1. Select stock as per drawing.
2. Place stock in 4-jaw independent chuck with stock projecting 1½". Adjust jaws so stock runs true.*
3. Arrange belt for proper spindle speed.*
4. Set lathe tool for facing.*
5. Face end.*
6. Center end for drilling, with centering tool held in tool post.*
7. Drill hole in end of cap with drill held in drill chuck in tailstock spindle. (See drawing.)
8. Set lathe tool for turning.*
9. Rough turn large diameter section.*
10. Rough turn small diameter section.
11. Finish turn and knurl large diameter section.*
12. Finish turn small diameter section.
13. Set tool for facing.*
14. Face shoulder of large diameter section.
15. Set tool for recessing.*
16. Cut recess for threading.
17. Arrange gearing for cutting thread as per drawing.*
18. Set tool for thread cutting.*
19. Cut thread on small diameter section to fit into threaded hole in large end of Part No. 2—Handle.*
20. Set tool for chamfering.
21. Chamfer end of thread as per drawing.
22. Remove drill chuck and place hardened center in tailstock spindle.*
23. Screw Part No. 2—Handle into place on thread of cap. Adjust tailstock to support small end of handle.*
24. File machined surfaces of handle and cap to remove tool marks. Polish with emery cloth.*
25. Withdraw tailstock. Unscrew and remove handle from lathe.
26. Set right hand tool for cutting off.*
27. Cut off cap to length as per drawing, plus ¹⁄₃₂" for finishing the end. Put oil on work and feed tool slowly by hand.

*See book "How to Run a Lathe."

MACHINIST'S HAMMER KIT (Continued)

OPERATIONS:

PART No. 3—CAP (Continued)

28. Screw cap into handle. Place handle in lathe chuck with cap end projecting ½", and true as before. Wrap emery cloth around handle to prevent marring finished surface.
29. Set lathe tool for facing.*
30. Face end of cap to length as per drawing.
31. Round end of cap with forming tool.
32. File end of cap to remove tool marks. Polish with emery cloth.*
33. Transfer to milling machine.
34. Saw screw driver slot in end of cap as per drawing. (See footnote C.)

PART No. 4—COLD CHISEL

1. Select stock as per drawing.
2. Forge end to shape as per drawing.
3. Grind and file to smooth finish.
4. Transfer to lathe.
5. Place stock in 4-jaw independent chuck, with hexagon end projecting ¼". Adjust jaws so stock runs true.*
6. Set tool for facing.*
7. Face end to length as per drawing.
8. Set tool for chamfering.
9. Chamfer corners as per drawing.
10. Transfer to forge. Harden point for 1 inch back from end. Draw temper as per drawing.*
11. Grind cutting edge to proper angle as per drawing.

PART No. 5—CENTER PUNCH

1. Select stock as per drawing.
2. Place stock in 4-jaw independent chuck projecting 2 inches. Adjust jaws so stock runs true.*
3. Arrange belt for proper spindle speed.*
4. Set compound rest at proper angle to obtain taper of 2 inches per foot. This is approximately 95 degrees to the right.
5. Set tool for taper turning.*
6. Rough turn taper, using round nose tool and feeding toward headstock with compound rest screw.*
7. Finish turn taper using fine feed for smooth finish.
8. Set compound rest at an angle of 120 degrees to the right. Turn angle of point using same tool.
9. File machined section lightly to remove tool marks. Polish with fine emery cloth.*
10. Turn work end for end in chuck with work projecting 1 inch, and true as before.
11. Set lathe tool for facing.*
12. Face end to length as per drawing.
13. Set tool for chamfering.
14. Chamfer corners as per drawing.
15. Transfer to forge, harden for ½" back from point and draw temper as per drawing.*
16. Place back in lathe chuck and re-polish.

PART No. 6—SCRIBE

1. Select stock as per drawing.
2. Place stock in 3-jaw universal chuck projecting 1¾" and running true.*
3. Arrange belt for proper spindle speed.*
4. Set compound rest at proper angle to obtain taper of 1¼" per foot. This is approximately 93 degrees to the right.
5. Set tool for taper turning.*
6. Rough turn taper using round nose tool and feeding toward headstock with compound rest screw.*

   NOTE A: Drilling can be done in lathe, using crotch center in tailstock spindle and drill held in drill chuck in headstock spindle.

   NOTE B: A relieved center would simplify the facing operation.
7. Finish turn taper to 1/32" in diameter at point, using fine feed for smooth finish. File to sharp point using fine file.
8. Turn work end for end in chuck, projecting ½" and true as before.
9. Set tool for facing.*
10. Face end to length as per drawing.*
11. Round end with a file. Polish with emery cloth.*
12. Transfer to forge. Harden point for 1" back from end. Draw temper as per drawing.*
13. Place back in lathe chuck and polish.

    NOTE C: If milling machine is not available, transfer work to bench and saw slot with a hack saw.

*See book "How to Run a Lathe."

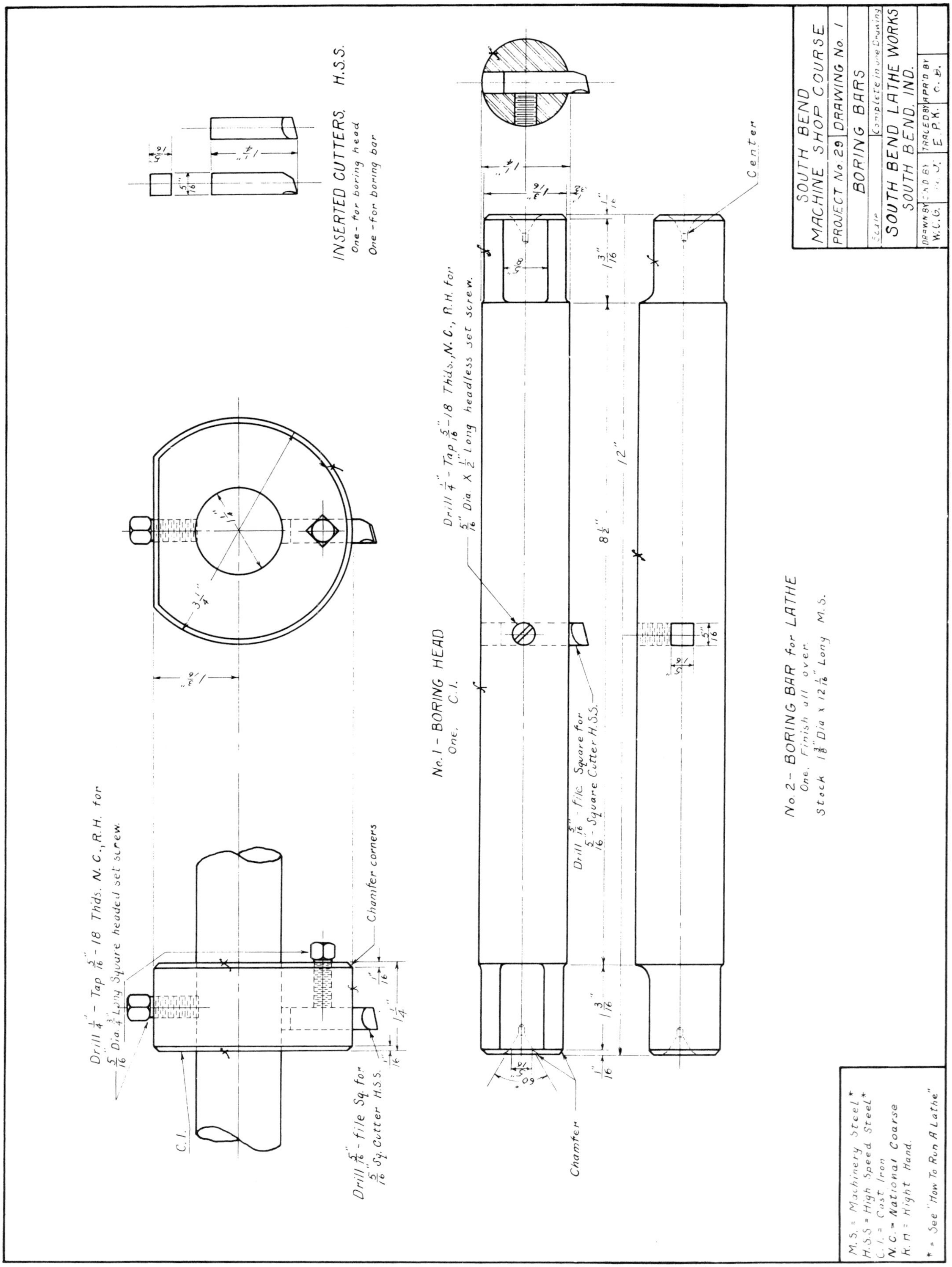
SOUTH BEND
MACHINE SHOP COURSE
PROJECT No. 29
DRAWING No. 1
BORING BARS
Scale
Complete in one Drawing
SOUTH BEND LATHE WORKS
SOUTH BEND, IND.
DRAWN BY W.C.G.
TRACED BY E.P.K.
APP'D BY C.B.
INSERTED CUTTERS, H.S.S.
One - for boring head
One - for boring bar
Center
Drill 1/4" - Tap 5/16"-18 Thds., N.C., R.H. for
5/16" Dia X 1/2" Long headless set screw.
No. 1 - BORING HEAD
One, C.I.
Drill 5/16" - file Square for
5/16" - Square Cutter H.S.S.
No. 2 - BORING BAR for LATHE
One, Finish all over
Stock 1 3/8" Dia x 12 1/16" Long M.S.
Drill 1/4" - Tap 5/16"-18 Thds. N.C., R.H. for
5/16" Dia 3/4" Long Square headed set screw.
Chamfer corners
C.I.
Drill 5/16" - file Sq. for
5/16" Sq. Cutter H.S.S.
Chamfer
M.S. = Machinery Steel*
H.S.S = High Speed Steel*
C.I. = Cast Iron
N.C. = National Coarse
R.H. = Right Hand.
* = See "How To Run A Lathe"

BORING BAR FOR THE LATHE

MATERIAL REQUIRED:

Part No. 1—Boring Head. Cast iron. One required.
Part No. 2—Boring Bar. Machinery steel 1⅜″ dia. x 12$\frac{1}{16}$″ long. One required.
Square head set screws $\frac{5}{16}$″ dia. x ¾″ long. Two required.
Inserted cutters, high speed tool steel $\frac{5}{16}$″ sq. x 1¼″ long. Two required.
Headless set screw $\frac{5}{16}$″ dia. x ½″ long. One required.

OPERATIONS:

PART No. 1—BORING HEAD

1. Remove sand and rough projections from casting.
2. Place casting in 4-jaw Independent Chuck with casting projecting ⅛″.*
3. Rough face side of casting.
4. Chamfer cored hole for starting the drill.*
5. Drill through cored hole with 3 or 4 lip drill.* Allow $\frac{1}{16}$″ stock for boring and reaming.
6. Bore hole allowing .005″ to .010″ for reaming hole to size.
7. Ream hole to size as per drawing.*
8. Transfer work to arbor press. Press mandrel into casting. (See footnote A.)
9. Place mandrel on centers in lathe,* with unfinished side of casting toward tailstock.
10. Rough face unfinished side of casting.*
11. Rough turn diameter.*
12. Finish turn diameter as per drawing.*
13. Finish face sides of casting to width as per drawing.
14. Chamfer outside corners as per drawing.
15. Transfer to bench. Lay off and center punch to drill holes for set screws and inserted tool steel cutter.
16. Transfer to drill press. Drill holes for set screws and inserted cutter as per drawing. (See footnote B.)
17. Transfer work to vise. Tap holes for set screws as per drawing.
18. File square hole to fit inserted tool steel cutter. Use a small square file.

PART No. 2—BORING BAR

1. Select stock as per drawing. (See footnote C.)
2. Lay off and center ends.*
3. Place stock on centers in lathe. (See footnote D.)
4. Face ends of length as per drawing.*
5. Chamfer ends of center holes as per drawing.
6. Rough turn diameter as far as lathe dog will permit.*
7. Turn stock end for end and rough turn remainder of diameter.
8. Transfer work to bench. Lay off and center punch to drill hole for inserted tool steel cutter and for set screw to hold cutter in place.
9. Transfer to drill press. Drill holes for inserted tool steel cutter, and for set screw for holding the cutter in place. (See footnote B.)
10. Transfer to shaper. Plane flat surfaces on ends of bar as per drawing. (See footnote E.)
11. Transfer work to bench. Tap hole for set screw to hold cutter in place.
12. File hole square to fit inserted cutter.
13. Transfer work back to lathe.
14. Undercut ends for dog as per drawing. Chamfer corners as per drawing.
15. Finish turn diameter allowing .003″ for filing to size.
16. File diameter to fit reamed hole in Part No. 1 —Boring Head. Polish with emery cloth.*

NOTE A: Remove dirt and put oil in hole before pressing mandrel in casting. Also clean and oil the mandrel.*

NOTE B: Drilling can be done in lathe, using drill pad and crotch center in tailstock spindle, and drill held in drill chuck in headstock spindle.*

NOTE C: If a large quantity of heavy work is to be done, the bar should be made of tool steel.

NOTE D: A relieved center would simplify the facing operation.*

NOTE E: If shaper is not available, remove stock by grinding and filing.

*See book "How to Run a Lathe."

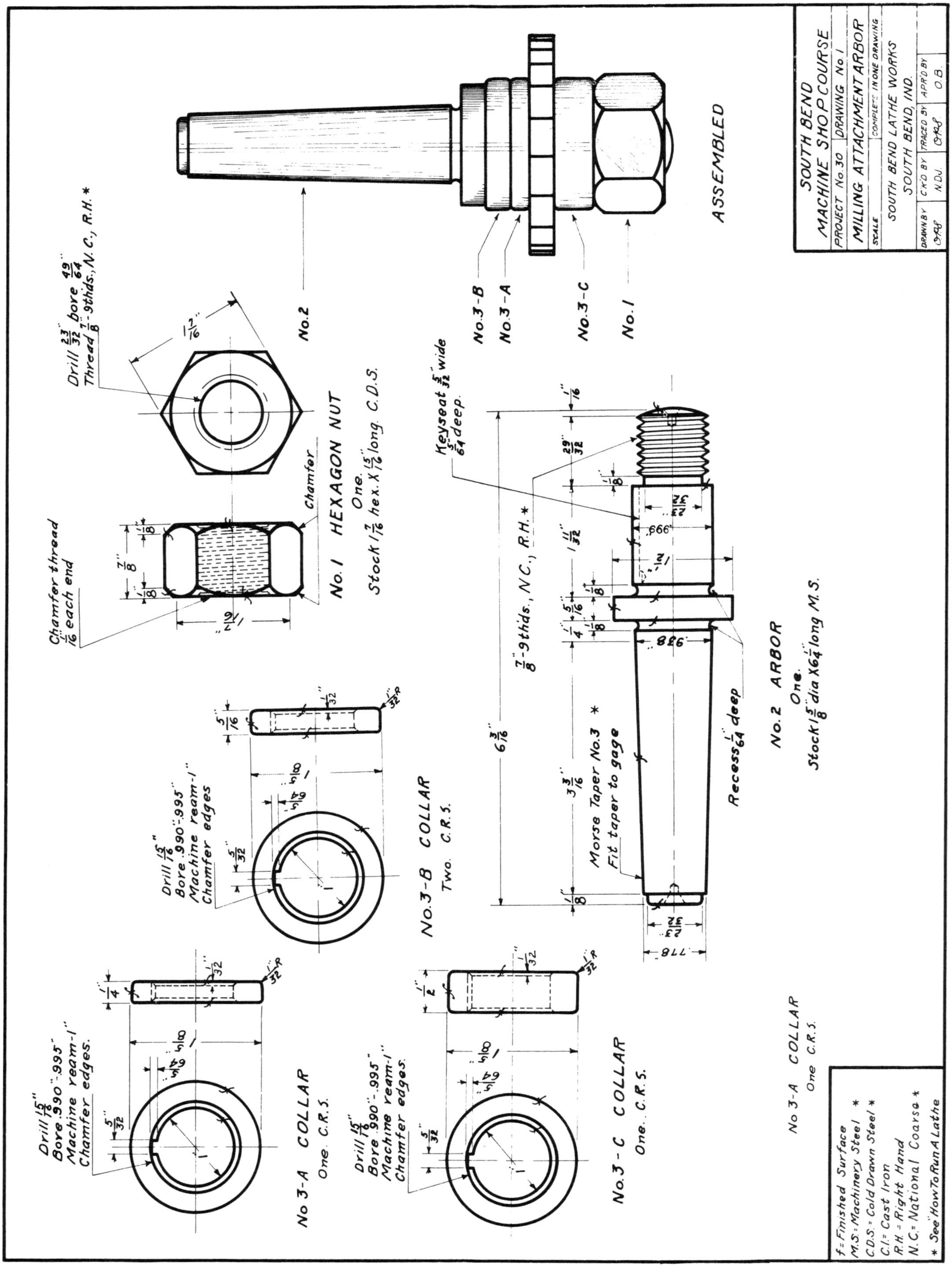
SOUTH BEND
MACHINE SHOP COURSE
PROJECT No. 30
DRAWING No. 1
MILLING ATTACHMENT ARBOR
SCALE
COMPLETE IN ONE DRAWING
SOUTH BEND LATHE WORKS
SOUTH BEND, IND.
DRAWN BY
CK'D BY N.D.J.
TRACED BY
APP'D BY O.B.
ASSEMBLED
No. 2
No. 3-B
No. 3-A
No. 3-C
No. 1
Drill 23/32" bore 49/64"
Thread 7/8"-9thds., N.C., R.H. *
1 7/16"
Chamfer
Chamfer thread 1/16" each end
No. 1 HEXAGON NUT
One.
Stock 1 7/16 hex. X 15/16 long. C.D.S.
Keyseat 5/32" wide 5/64" deep.
7/8"-9thds., N.C., R.H. *
.998"
.938"
Morse Taper No. 3 *
Fit taper to gage
Recess 1/64" deep
.778"
6 3/16"
3 3/16"
No. 2 ARBOR
One.
Stock 1 5/8" dia X 6 1/4" long M.S.
Drill 15/16"
Bore .990"-.995"
Machine ream-1"
Chamfer edges
No. 3-B COLLAR
Two. C.R.S.
No 3-A COLLAR
One. C.R.S.
No. 3-C COLLAR
One. C.R.S.
f = Finished Surface
M.S. = Machinery Steel *
C.D.S. = Cold Drawn Steel *
C.I. = Cast Iron
R.H. = Right Hand
N.C. = National Coarse *
* See "How To Run A Lathe"

MILLING ARBOR FOR LATHE

MATERIAL REQUIRED:

Part No. 1—Nut. Cold drawn steel 1 7/16" hex. x 15/16" long. One required.
Part No. 2—Arbor. Machinery steel 1 5/8" dia. x 6 1/4" long. One required.
Part No. 3A—Collar. Cast iron. One required.
Part No. 3B—Collar. Cast iron. Two required.
Part No. 3C—Collar. Cast iron. One required.

OPERATIONS:

PART No. 1—HEXAGON NUT

1. Select stock as per drawing.
2. Place stock in a three-jaw chuck with stock projecting 1/8".*
3. Face end.*
4. Center stock for drilling.*
5. Drill hole through stock as per drawing.
6. Bore hole to size as per drawing.*
7. Arrange gears for cutting thread as per drawing.*
8. Set tool for internal thread cutting.*
9. Cut thread, leaving .005" stock for tap to remove.*
10. Finish thread to size with tap before removing work from lathe.*
11. Chamfer end of thread as per drawing.
12. Chamfer outside corners as per drawing.
13. Turn work end for end in chuck and face end to length as per drawing.
14. Chamfer end of thread as per drawing.
15. Chamfer outside corners as per drawing.
16. Transfer to forge. Caseharden all over.*

PART No. 2—ARBOR

1. Select stock as per drawing.
2. Lay off and center ends.*
3. Place stock on centers in lathe.* (See footnote A.)
4. Face ends to length as per drawing.
5. Rough turn 1" diameter section.*
6. Rough turn large diameter section.
7. Rough turn section to be threaded.
8. Rough turn taper section to approximately 1 1/32" in diameter.
9. Arrange lathe for taper turning.*
10. Take light trial cut over tapered section.
11. Test the taper in No. 3 Morse taper hole.*
12. If taper is not correct, adjust setting to correct error.
13. Take second trial cut and test as before. Repeat until proper taper.
14. Finish turn tapered section, allowing .003" for filing to diameter as per drawing. (See footnote B.)
15. Arrange lathe for straight turning.*
16. Cut shoulder for clearance at small end of taper as per drawing. Round corner with file.
17. File tapered section to size as per drawing.*
18. Finish face shoulder of large diameter section.*
19. Turn stock end for end and finish face other shoulder of large diameter section.
20. Finish turn large diameter section.*
21. Finish turn 1" diameter section. Allow .003" for filing to diameter as per drawing. (See footnote B.)
22. Finish turn section to be threaded .010" undersize to insure free fit on thread.
23. Finish face shoulder of 1" diameter section.
24. Cut threading recess as per drawing.
25. Arrange gears for cutting thread as per drawing.*
26. Set tool for thread cutting.*
27. Cut thread as per drawing, fitting to Part No. 1—Nut.*
28. Round corners of large diameter section with file.
29. Round threaded end as per drawing, not touching center hole. Use forming tool.
30. File machined surfaces to size as per drawing. Polish with emery cloth.*
31. Transfer work to bench. Lay off and center punch to drill hole at end of keyseat as per drawing.
32. Transfer work to drill press. Drill hole at end of keyseat as per drawing.
33. Transfer work to shaper. Cut keyseat as per drawing.
34. File burr off edges of keyseat.

PARTS Nos. 3A, 3B, 3C—COLLARS

NOTE: Perform the following operations on all four collars.

1. Select casting.
2. Place casting in three-jaw chuck with casting projecting 1/8" and running true.*
3. Rough face end.*
4. Chamfer edge of cored hole for starting the drill.*
5. Drill through cored hole with 15/16" drill (3 or 4 lip).*
6. Bore hole leaving .005" to .010" for reaming hole to size.*
7. Ream hole to size as per drawing. Use machine reamer.*
8. Chamfer hole as per drawing.
9. Transfer to arbor press and press mandrel into work from finished side. (See footnote C.)
10. Place mandrel on centers in lathe with unfinished end toward tailstock.*
11. Rough turn outside diameter.*
12. Rough face unfinished side of collar.*
13. Use right and left-hand facing tools, and finish face both sides of collar without changing position of mandrel on the lathe centers.*
14. Remove work from mandrel, place in chuck and chamfer other side of hole as per drawing.
15. When all collars (4) have been finish faced, and chamfered, place them on milling attachment arbor, secure in position with nut, and place arbor on centers in lathe.
16. Finish turn outside diameter of collars.*
17. File lightly to remove tool marks. Polish with emery cloth.*
18. Remove collars from milling attachment arbor.
19. Place collars on lathe mandrel between centers, one at a time, and round outside corners with a file, as per drawing.
20. Transfer collars to shaper and keyseat as per drawing.

NOTE A: A relieved center would simplify the facing operation.*

NOTE B: If it is desired to finish by grinding, finish turn to .010" overside and cut recess for clearance for grinding as per drawing.

NOTE C: Remove dirt and put oil in hole before pressing arbor into casting. Also clean and oil the arbor.*

*See book "How to Run a Lathe."

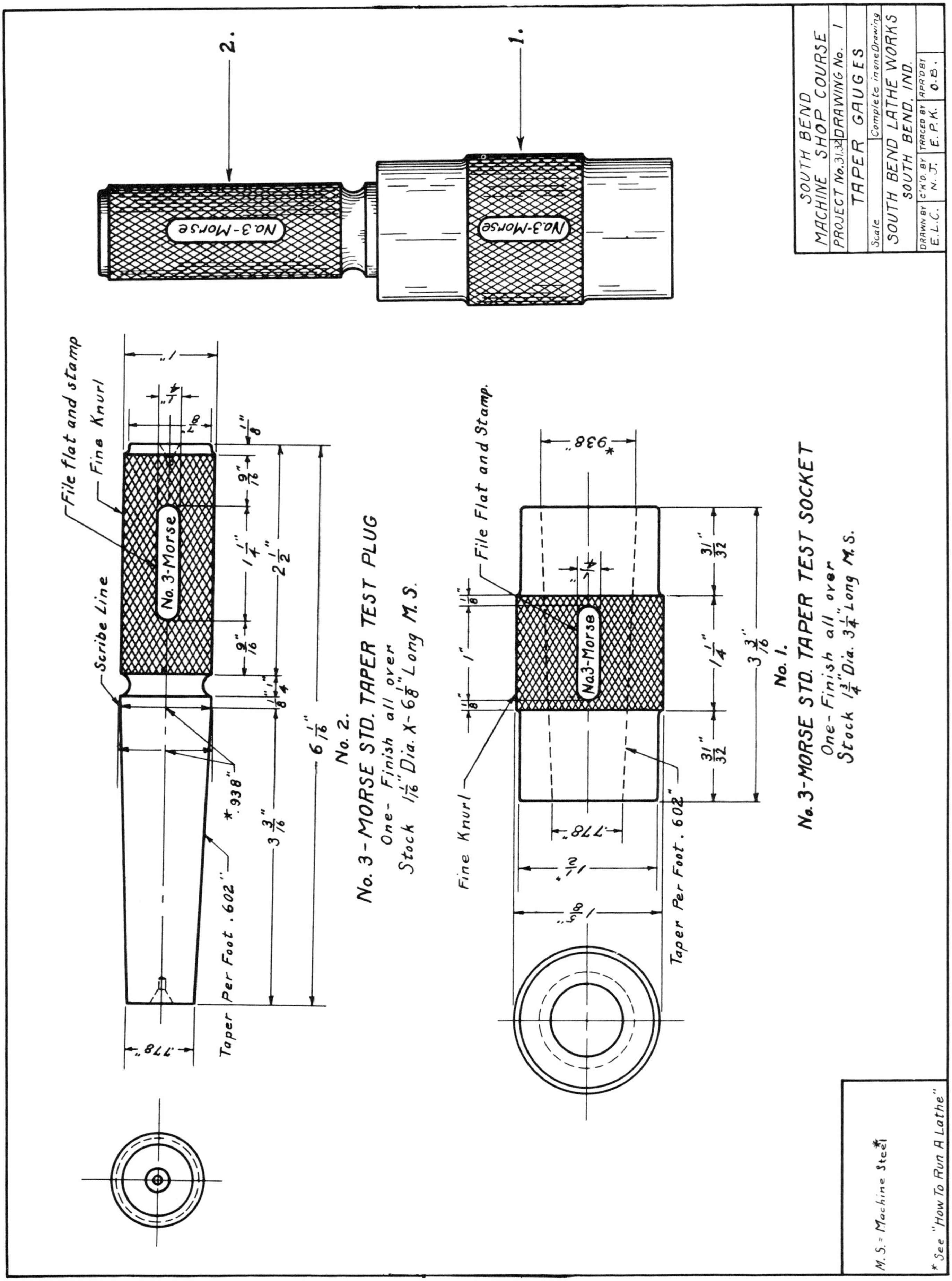
SOUTH BEND MACHINE SHOP COURSE
PROJECT No. 31,32 DRAWING No. 1
TAPER GAUGES
Scale
Complete in one Drawing
SOUTH BEND LATHE WORKS
SOUTH BEND, IND.
DRAWN BY E.L.C.
C'K'D. BY N.J.
TRACED BY E.P.K.
APP'D BY O.B.
2.
1.
No. 3-Morse
File flat and stamp
Fine Knurl
Scribe Line
Taper Per Foot .602"
No. 3-MORSE STD. TAPER TEST PLUG
One - Finish all over
Stock 1 1/16" Dia. X 6 5/8" Long M.S.
No. 2.
File Flat and Stamp.
Fine Knurl
No. 3-MORSE STD. TAPER TEST SOCKET
One - Finish all over
Stock 1 3/4" Dia. 3 1/4" Long M.S.
No. 1.
M.S. = Machine Steel
* See "How To Run A Lathe"

MORSE TAPER STANDARD TEST PLUG

MATERIAL REQUIRED:

Machine Steel 1 1/16" dia. x 6 1/8" long. One required.

OPERATIONS:

1. Select stock as per drawing.
2. Lay off and center ends.*
3. Place stock on centers in lathe.*
4. Face ends to length as per drawing.
5. Rough turn straight diameter.
6. Finish turn straight diameter.
7. Knurl straight diameter as per drawing.*
8. Turn recess between knurled and taper sections.
9. Undercut end as per drawing. Use same tool. Round corner with file.
10. Arrange lathe for turning taper as per drawing.*
11. Take light trial cut over taper section.
12. Test taper in taper hole in No. 3 Morse taper standard test socket. If taper is not correct, adjust setting to correct error.
13. Take second trial cut and test as before. Repeat until taper is correct.
14. Rough turn taper section.
15. Finish turn taper, allowing .003" for filing to size as per drawing. (See footnote A.)
16. File tapered section to size as per drawing, fitting to test socket.
17. Scribe line 1/8" from large end of taper section as per drawing. Use tool ground to sharp point.
18. Polish with emery cloth.*
19. Transfer to vise. File flat on surface on knurled section, and stamp test plug as per drawing. Polish surface with emery cloth.

NOTE A: If it is desired to finish taper by grinding allow 1/32" for finishing, and harden tapered section. Draw temper to dark straw color.*

PROJECT No. 32

MORSE TAPER STANDARD TEST SOCKET

MATERIAL REQUIRED:

Machine steel 1 3/4" dia. x 3 1/4" long. One required.

OPERATIONS:

1. Select stock as per drawing.
2. Place stock in 4-jaw Independent Chuck with stock projecting 1/4".
3. Face end.*
4. Center end of stock for drilling.*
5. Drill through stock with 23/32" drill.*
6. Arrange lathe for boring taper as per drawing.*
7. Set small tool for taper boring.*
8. Take light trial cut in taper hole.
9. Test taper with a No. 3 Morse taper reamer.*
10. If the taper is not correct, adjust the setting to correct error. Repeat until taper is correct.
11. Rough bore taper hole.
12. Finish bore taper hole as per drawing. Allow .008" for reaming.
13. Arrange lathe for reaming the taper hole.
14. Throw in back gears. Adjust belt to large step of spindle cone.*
15. Start lathe and feed reamer into work slowly with tailstock handwheel. Remove reamer and clean chips out of flutes often to prevent clogging and tearing the surface of the hole.
16. Ream taper hole to within .003" of size as per drawing.

    NOTE A: Do not clamp work in vise. Clamp reamer in vise on square end of shank. Put a common lathe dog on work to use as a wrench. Place work on reamer and ream to size very carefully by hand. Put oil on reamer. The reamer enlarges the hole approximately .005" in advancing 1/8" into work.

17. Transfer work to bench vise. Finish reaming taper hole by hand. (See footnote A.)
18. Throw out back gears. Arrange belt for proper spindle speed for turning.*
19. Select a piece of steel, center and face the ends, and make a temporary arbor for turning the outside of work. Fit arbor into taper hole in work. (See footnote B).
20. Arrange lathe for straight turning.*
21. Press arbor in work and place arbor on centers in lathe* with unfinished end of work toward the tailstock. (See footnote C.)
22. Face unfinished end of stock to length as per drawing.
23. Rough turn small diameter sections.*
24. Rough turn center section.
25. Finish turn center section as per drawing.
26. Knurl center section as per drawing.
27. Finish turn small diameter sections using tool ground to form fillet at shoulder.
28. File machined surfaces, round corners and polish small diameter sections.*
29. Transfer work to bench vise. File flat surface on knurled section. Stamp test socket as per drawing. Polish surface with emery cloth.

NOTE B: The stock for the arbor should be 1/8" larger in diameter than the large end of taper hole and 2" longer than work.

NOTE C: Remove dirt and put oil in taper hole before putting arbor in work. Also clean and oil the arbor.

*See book "How to Run a Lathe."

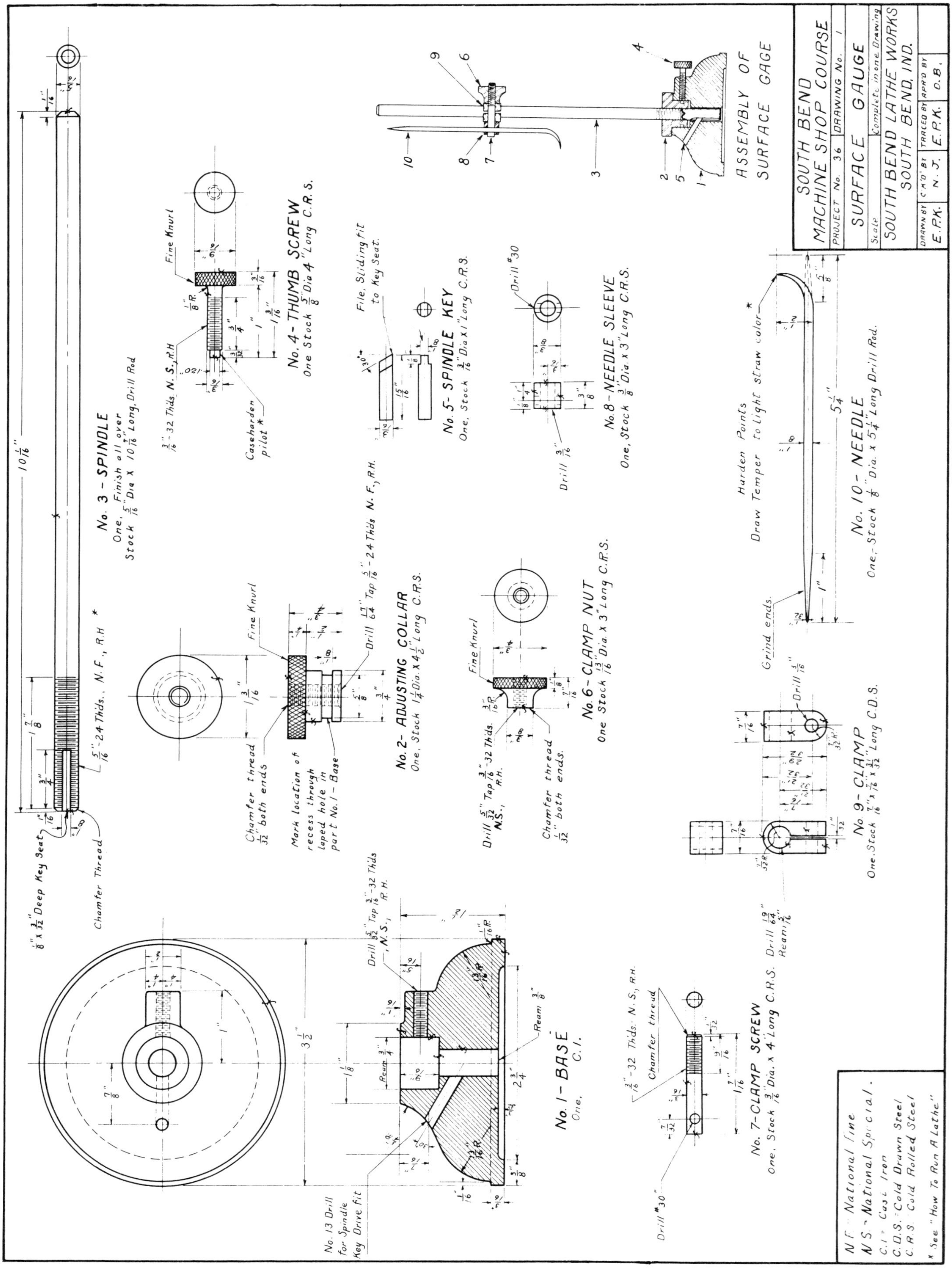
SOUTH BEND
MACHINE SHOP COURSE
PROJECT No. 36 DRAWING No. 1
SURFACE GAUGE
Scale Complete in one Drawing
SOUTH BEND LATHE WORKS
SOUTH BEND, IND.
DRAWN BY E.P.K. CH'K'D BY N.J. TRACED BY E.P.K. APP'D BY O.B.
ASSEMBLY OF SURFACE GAGE
No. 1 - BASE C.I. One,
No. 2 - ADJUSTING COLLAR One, Stock 1 1/4" Dia. X 4 1/2" Long C.R.S.
No. 3 - SPINDLE One, Finish all over Stock 5/16" Dia X 10 7/16" Long, Drill Rod
No. 4 - THUMB SCREW One Stock 5/8" Dia 4" Long C.R.S.
No. 5 - SPINDLE KEY One, Stock 3/16" Dia X 1" Long C.R.S.
No. 6 - CLAMP NUT One Stock 13/16" Dia X 3" Long C.R.S.
No. 7 - CLAMP SCREW One, Stock 3/16" Dia. X 4" Long C.R.S.
No. 8 - NEEDLE SLEEVE One, Stock 3/8" Dia. X 3" Long C.R.S.
No. 9 - CLAMP One, Stock 7/16" X 7/16" X 31/32" Long C.D.S.
No. 10 - NEEDLE One, Stock 1/8" Dia. X 5 1/4" Long Drill Rod.
Harden Points Draw Temper to light straw color *
Grind ends.
Fine Knurl
Casehardened pilot *
Chamfer thread 1/32" both ends
Mark location of recess through laped hole in part No. 1 - Base
File, Sliding fit to Key Seat.
Chamfer Thread
5/16" - 24 Thds. N.F., R.H. *
No. 13 Drill for Spindle Key Drive fit
Drill #30
N.F. - National Fine
N.S. - National Special.
C.I. - Cast Iron
C.D.S. - Cold Drawn Steel
C.R.S. - Cold Rolled Steel
* See "How To Run A Lathe"

SURFACE GAUGE

MATERIAL REQUIRED:

Part No. 1—Base. Cast iron. One required.
Part No. 2—Adjusting collar. Cold rolled steel 1¼″ dia. x 4½″ long. One required.
Part No. 3—Spindle. Drill rod 5/16″ dia. x 10 7/16″ long. One required.
Part No. 4—Thumb Screw. Cold rolled steel ⅝″ dia. x 4″ long. One required.
Part No. 5—Spindle Key. Cold rolled steel 3/16″ dia. x 1″ long. One required.
Part No. 6—Clamp Nut. Cold rolled steel 13/16″ dia. x 3″ long. One required.
Part No. 7—Clamp Screw. Cold rolled steel ¼″ dia. x 4″ long. One required.
Part No. 8—Needle Sleeve. Cold rolled steel 7/16″ x 3″ long. One required.
Part No. 9—Clamp. Cold drawn steel 7/16″ sq. x 31/32″ long. One required.
Part No. 10—Needle. Drill rod ⅛″ dia. x 5¼″ long. One required.

OPERATIONS:

PART No. 1—BASE

1. Remove sand and rough projections from casting.
2. Place casting in 4-jaw Independent Chuck* with flange end of casting projecting ⅜″.
3. Rough face flange side of casting.*
4. Rough turn diameter of flange.*
5. Turn work end for end in chuck. Rough and finish face other side of casting.
6. Center casting for drilling.*
7. Drill through casting with 5/16″ drill.
8. Bore hole allowing .005″ to .010″ stock for reaming to size.
9. Ream hole to size as per drawing.*
10. Use boring tool and counterbore hole as per drawing. Allow .005″ to .010″ for reaming to size.
11. Ream hole to size as per drawing.*
12. Use boring tool and square bottom of hole as per drawing.
13. Transfer work to arbor press. Press a mandrel in ⅜″ hole from counterbored end of casting. (See footnote A.)
14. Place mandrel on centers in lathe.* Finish face flange side of casting to height as per drawing.*
15. Finish turn diameter of flange as per drawing.*
16. File corners to remove sharp edges. File and polish diameter of flange.*
17. Transfer work to bench. Lay off and center-punch to drill holes for thumb screw and spindle key.
18. Transfer work to drill press. Drill holes for thumb screw and spindle key as per drawing.
19. Transfer work to bench. Tap hole for thumb screw as per drawing.
20. Paint unfinished part of casting.

PART No. 2—ADJUSTING COLLAR

1. Select stock as per drawing.
2. Place stock in 3-jaw Universal Chuck* with stock projecting 1½″.
3. Face and center drill end*, and adjust tailstock to support end of work.*
4. Rough turn large diameter section.*
5. Rough turn small diameter section. Rough face shoulder, using same tool.
6. Finish turn large diameter section as per drawing.*
7. Knurl large diameter section as per drawing.*
8. Withdraw tailstock. Center end of stock for drilling.*
9. Drill into end of work 2″ deep with 17/64″ drill.*
10. Start tap into work to depth of ½″.*
11. Back tap out by hand. Chamfer end of thread as per drawing.
12. Finish turn small diameter section, allowing .003″ for filing to diameter.
13. Finish face shoulder to length as per drawing. File corners to remove sharp edges.
14. File small diameter section to size*, fitting to ¾″ reamed hole in Part No. 1—Base.
15. Place Part No. 1—Base on small diameter section and against the shoulder. Use a scriber and mark location of recess in small diameter section, through the tapped hole in Part No. 1—Base.
16. Cut recess in small diameter section in the location just marked, and to depth as per drawing.
17. Use right-hand tool* and cut off work to length as per drawing, plus 1/32″ stock for finishing the end.
18. Transfer work to bench vise. Finish tapping hole by hand.
19. Transfer work to lathe. Place work in 3-jaw chuck with knurled end projecting.
20. Face end to length as per drawing. File corner to remove sharp edge. Polish the face.*
21. Chamfer end of thread as before.

*See book "How to Run a Lathe."

SURFACE GAUGE (Continued)

OPERATIONS:

PART No. 3—SPINDLE

1. Select stock as per drawing.
2. Place stock in 3-jaw Universal Chuck* with end of stock projecting ½". It is important that the stock run true.*
3. Face end.* Center end of stock for drilling.* Center drill end using center drill with 1/16" pilot drill.
4. Turn stock end for end in chuck. Face end to length as per drawing, plus 5/16" stock for removing one of the center holes. Center stock for drilling and center drill end as before.
5. Place stock on centers in lathe.*
6. Arrange gearing for cutting thread as per drawing.*
7. Set lathe tool for thread cutting.*
8. Cut thread as per drawing.* Fit into tapped hole in Part No. 2—Adjusting Collar.
9. Chamfer end of thread as per drawing.
10. Polish spindle all over.*
11. Transfer work to milling machine. Mill keyseat as per drawing.
12. Place work in 3-jaw Universal Chuck with unfinished end projecting ½".
13. Face end to length as per drawing, removing the center hole.
14. Chamfer corner as per drawing. File and polish the end.*

PART No. 4—THUMB SCREW

1. Select stock as per drawing.
2. Place stock in 3-jaw Universal Chuck* with stock projecting 2".
3. Face end.* Center end of stock for drilling.* Center drill end, using center drill with a 1/16" pilot drill.
4. Adjust tailstock to support end of work.*
5. Rough turn large diameter section.*
6. Rough turn small diameter section to ½" in diameter, allowing ¼" stock for removing the center hole.
7. Finish turn large diameter section as per drawing.
8. Knurl large diameter section as per drawing.*
9. Rough turn small diameter section to approximate diameter. Rough turn fillet and rough face shoulder.
10. Finish turn small diameter section. Use tool ground to form fillet at shoulder.*
11. Finish face shoulder to length as per drawing, using same tool. File corner to remove sharp edge.
12. Arrange gearing for cutting thread as per drawing.*
13. Set lathe tool for thread cutting.*
14. Cut thread as per drawing*, fitting to 3/16" tapped hole in Part No. 1—Base.
15. Turn pilot section on threaded end as per drawing, allowing ¼" for removing the center hole.
16. Withdraw tailstock and (with a hack saw) saw off 7/32" from pilot section to remove center hole.
17. Face end to length as per drawing. File corner to remove sharp edge.
18. Use right-hand tool* and cut off work to length as per drawing, plus 1/32" stock for facing the end.
19. Place work in 3-jaw Universal Chuck, with knurled end projecting. Grip work lightly to prevent marring thread.
20. Face end to length as per drawing. File corner to remove sharp edge.
21. Transfer work to forge. Caseharden pilot section.*

PART No. 5—SPINDLE KEY

1. Select stock as per drawing.
2. File end to approximate angle as per drawing.
3. File flat surfaces on sides at end of stock as per drawing, allowing 1/64" for fitting.
4. Oil and drive key into No. 13 drill hole in Part No. 1—Base.
5. Use a knife file and dress key for sliding fit in keyway of Part No. 3—Spindle.
6. File outside end of key flush with surface of casting.

PART No. 6—CLAMP NUT

1. Select stock as per drawing.
2. Place stock in 3-jaw Universal Chuck* with stock projecting 1¼".
3. Face end.*
4. Rough turn large diameter section.*
5. Rough turn small diameter section. Rough turn fillet and rough face shoulder.
6. Finish turn large diameter section as per drawing.
7. Knurl large diameter section as per drawing.*
8. Finish turn small diameter section, using tool ground to form fillet at shoulder.
9. Finish face shoulder, using same tool.

*See book "How to Run a Lathe."

SURFACE GAUGE (Continued)

OPERATIONS:

PART No. 6—CLAMP NUT (Continued)

10. File corners to remove sharp edges. Polish machined surface.*
11. Use right-hand tool* and cut off work to length as per drawing, plus 1/32″ stock for finishing the end.

NOTE: Stop work on Part No. 6—Clamp Nut. Take up and complete Part No. 7—Clamp Screw. Then use Part No. 7—Clamp Screw as a stub arbor to face unfinished end of Part No. 6—Clamp Nut.

PART No. 7—CLAMP SCREW

1. Select stock as per drawing.
2. Place stock in 3-jaw Universal Chuck* with stock projecting 1¾″. It is important that the stock run true.*
3. Face end.* Center end of stock for drilling.* Center drill end, using center drill with a 1/16″ pilot drill.
4. Adjust tailstock to support end of work.*
5. Turn diameter as per drawing.
6. Arrange gearing for cutting thread as per drawing.*
7. Set lathe tool for thread cutting.*
8. Cut thread,* fitting to tapped hole in Part No. 6—Clamp Nut.
9. Chamfer end of thread as per drawing.
10. Use right-hand tool* and cut off work to length as per drawing, plus 1/32″ stock for finishing the end.
11. Place work in 3-jaw chuck with unfinished end projecting ⅛″.
12. Face end to length as per drawing.
    NOTE: Do not drill hole through diameter until instructed.
13. Place Part No. 7—Clamp Screw in 3-jaw chuck with threaded end projecting ⅜″.
14. Screw Part No. 6—Clamp Nut on end of Part No. 7—Clamp Screw, with finished end against face of chuck jaws.
15. Face unfinished end of Part No. 6—Clamp Nut to length as per drawing. File corner to remove sharp edge. Polish the face.*

PART No. 8—NEEDLE SLEEVE

1. Select stock as per drawing.
2. Place stock in 3-jaw Universal Chuck* with stock projecting ¾″.
3. Face end.*
4. Center end of stock for drilling.*
5. Drill into end of stock ½″ deep with 3/16″ drill.*
6. Rough turn diameter.*
7. Finish turn diameter.*
8. Use right-hand tool* and cut off work to length as per drawing, plus 1/32″ stock for finishing the end.
9. Place work in 3-jaw chuck with unfinished end projecting 1/16″.
10. Face end to length as per drawing.
11. Transfer work to bench. Lay off and center punch to drill hole through diameter.
12. Place Part No. 7—Clamp Screw, and Part No. 8—Needle Sleeve together in position they will occupy when assembled. (See assembly drawing.)
13. Transfer work to drill press. Clamp the parts firmly in position and drill hole through diameter, as per drawing, drilling through both parts.

PART No. 9—CLAMP

1. Select stock as per drawing.
2. Place stock in 4-jaw Independent Chuck* with stock projecting ⅛″.
3. Face ends to length as per drawing.
4. Transfer work to bench. Lay off and center punch to drill and ream as per drawing. Lay off rounded ends as per drawing.
5. Transfer work to drill press. Drill both holes as per drawing. Ream large hole, using machine reamer.
6. Transfer work to emery wheel. Rough grind rounded ends.
7. Transfer work to bench vise. Finish rounded ends to shape with a file.
8. Transfer to milling machine and saw slot as per drawing. (See footnote B.)
9. Transfer work to bench vise. Remove all burrs and sharp corners with a fine file. Polish all over.*

PART No. 10—NEEDLE

1. Select stock as per drawing.
2. Grind ends to sharp point as per drawing.
3. Transfer work to forge. Bend hook end as per drawing. Harden points and draw temper as per drawing.*

NOTE A: Remove dirt and put oil in hole before pressing mandrel into casting.* Also clean and oil the mandrel.*

NOTE B: If milling machine is not available, saw slot with a hack saw.

*See book "How to Run a Lathe."

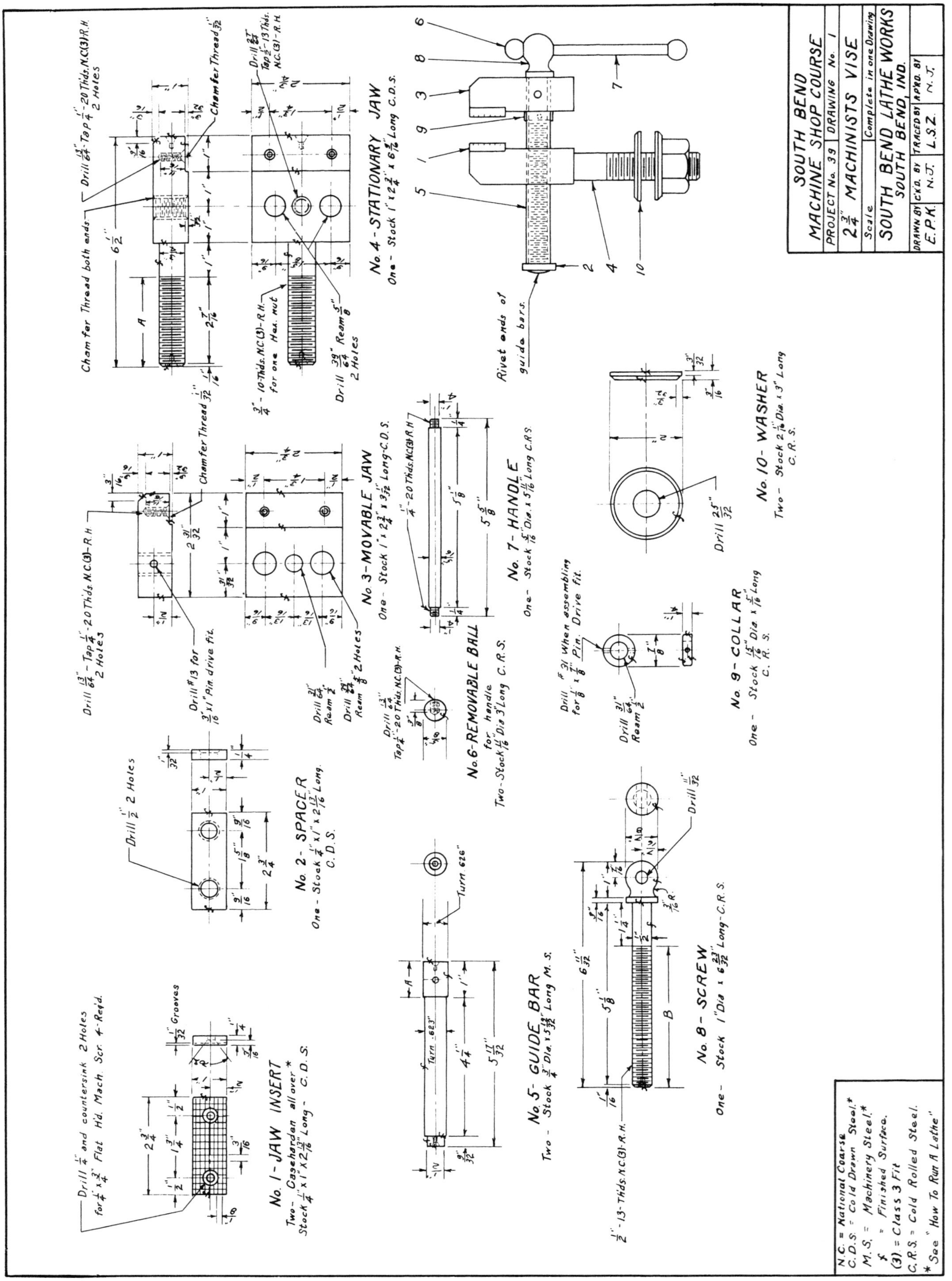
SOUTH BEND
MACHINE SHOP COURSE
PROJECT No. 39 DRAWING No. 1
2 3/4" MACHINISTS VISE
Scale
Complete in one Drawing
SOUTH BEND LATHE WORKS
SOUTH BEND, IND.
DRAWN BY E.P.K.
CK'D. BY N.J.
TRACED BY L.S.Z.
APRD. BY N.J.
Rivet ends of guide bars.
Drill 13/64" Tap 1/4"-20 Thds. N.C.(3) R.H. 2 Holes
Chamfer Thread both ends
Chamfer Thread 1/32"
Drill 27/64" Tap 1/2"-13 Thds. N.C.(3)-R.H.
3/4"-10 Thds. N.C.(3)-R.H. for one Hex. nut
Drill 39/64" Reem 5/8" 2 Holes
No. 4-STATIONARY JAW
One- Stock 1" x 2 1/4" x 6 7/16" Long C.D.S.
Drill 13/64" - Tap 1/4"-20 Thds. N.C.(3)-R.H. 2 Holes
Drill #13 for 3/16" x 1" Pin drive fit.
Drill 31/64" Ream 1/2"
Drill 39/64" Ream 5/8" 2 Holes
No. 3-MOVABLE JAW
One- Stock 1" x 2 1/4" x 3 1/32" Long-C.D.S.
1/4"-20 Thds. N.C.(3) R.H.
No. 7-HANDLE
One- Stock 5/16" Dia. x 5 11/16" Long C.R.S.
No. 10-WASHER
Two- Stock 2 1/16" Dia. x 3" Long C.R.S.
Drill 25/32"
Drill 13/64" Tap 1/4"-20 Thds. N.C.(3)-R.H.
No. 6-REMOVABLE BALL for handle
Two-Stock 11/16" Dia 3" Long C.R.S.
Drill #31 When assembling for 1/8" x 7/8" Pin. Drive fit.
Drill 31/64" Ream 1/2"
No. 9-COLLAR
One- Stock 15/16" Dia x 5/16" Long C.R.S.
Drill 1/2" 2 Holes
No. 2-SPACER
One- Stock 1/4" x 1" x 2 13/16" Long. C.D.S.
Turn .626"
Turn .623"
Drill 11/32"
Drill 1/4" and countersink 2 Holes for 1/4" x 3/4" Flat Hd. Mach. Scr. 4-Req'd.
Grooves
No. 1-JAW INSERT
Two- Casehardan all over *
Stock 1/4" x 1" x 2 13/16" Long - C.D.S.
No. 5-GUIDE BAR
Two - Stock 3/4" Dia. x 5 13/32" Long M.S.
1/2"-13 Thds. N.C.(3)-R.H.
No. 8-SCREW
One- Stock 1" Dia x 6 23/32" Long-C.R.S.
N.C. = National Coarse
C.D.S. = Cold Drawn Steel.*
M.S. = Machinery Steel.*
f = Finished Surface.
(3) = Class 3 Fit
C.R.S. = Cold Rolled Steel.
* See "How To Run A Lathe"

SMALL BENCH VISE

MATERIAL REQUIRED:

Part No. 1—Jaw Insert. Cold drawn steel $\frac{1}{4}''$ x $1''$ x $2\frac{13}{16}''$ long. Two required.
Part No. 2—Spacer. Cold drawn steel $\frac{1}{4}''$ x $1''$ x $2\frac{13}{16}''$ long. One required.
Part No. 3—Movable Jaw. Cold drawn steel $1''$ x $2\frac{3}{4}''$ x $3\frac{1}{32}''$ long. One required.
Part No. 4—Stationary Jaw. Cold drawn steel $1''$ x $2\frac{3}{4}''$ x $6\frac{9}{16}''$ long. One required.
Part No. 5—Guide Bar. Machinery steel $\frac{3}{4}''$ dia. x $5\frac{19}{32}''$ long. Two required.
Part No. 6—Removable Ball. Cold rolled steel $\frac{11}{16}''$ dia. x $3''$ long. One required.
Part No. 7—Handle. Cold rolled steel $\frac{5}{16}''$ dia. x $5\frac{11}{16}''$ long. One required.
Part No. 8—Screw. Cold rolled steel $1''$ dia. x $6\frac{23}{32}''$ long. One required.
Part No. 9—Collar. Cold rolled steel $\frac{15}{16}''$ dia. x $\frac{5}{16}''$ long. One required.
Part No. 10—Washer. Cold rolled steel $2\frac{1}{16}''$ dia. x $3''$ long. One required.
Flat Hd. Mach. Screws $\frac{1}{4}''$ dia. x $\frac{3}{4}''$ long. Four required.
Cold Rolled Steel Pins $\frac{3}{16}''$ dia. x $1''$ long. Two required.
Cold Rolled Steel Pin $\frac{1}{8}''$ dia. x $\frac{15}{16}''$ long. One required.
Hex. Nut $\frac{3}{4}''$ U. S. S.—R. H. One required.

OPERATIONS:

PART No. 1—JAW INSERT

1. Select two pieces of stock as per drawing. Perform following operations on both parts.
2. Finish ends to length by planing or grinding and filing.
3. Lay off, centerpunch, drill and countersink as per drawing. (See footnote A.)
4. Lay off and plane grooves as per drawing. (See footnote B.)
5. Caseharden all over.*

PART No. 2—SPACER

1. Select stock as per drawing.
2. Finish ends to length by planing or grinding and filing.
3. Lay off, centerpunch, drill holes and countersink as per drawing. (See footnote A.)

PART No. 3—MOVABLE JAW

1. Select stock as per drawing.
2. Finish ends to length by planing or grinding and filing.
3. Lay off location of shoulder next to offset surface.
4. Set work square in shaper vise. Plane offset surface and square shoulder as per drawing.
5. Turn stock over and chamfer outside corner as per drawing.
6. Lay off, centerpunch and drill $\frac{31}{64}''$ and $\frac{39}{64}''$ holes as per drawing. (See footnote A.) Do not ream until instructed.
7. Clamp Jaw Insert, Part No. 1 and Movable Jaw, Part No. 3, together as per assembly drawing.
8. Using Part No. 1 as a jig, spot location of screw holes on Part No. 3.
9. Drill screw holes in Part No. 3 as per drawing. (See footnote A.)
10. With the parts still clamped together, tap screw holes in Part No. 3 as per drawing, using holes in Part No. 1 to guide tap.
11. After both holes have been tapped and the screws fitted, Part No. 1 should be removed until the vise is assembled. (See footnote C.)
12. Chamfer thread as per drawing.
13. Remove burrs and sharp corners. File and polish all surfaces except the offset surface.*

PART No. 4—STATIONARY JAW

1. Select stock as per drawing.
2. Lay off and center ends.*
3. Face ends to length as per drawing.* (See footnote D.)
4. Rough turn round section and rough face shoulder.
5. Finish turn round section as per drawing.*
6. Turn thread section (A) .010″ undersize.
7. Face shoulder to length as per drawing.
8. Arrange gearing for cutting thread as per drawing.*
9. Set lathe tool for thread cutting.*

*See book "How to Run a Lathe."

SMALL BENCH VISE (Continued)

OPERATIONS:

PART No. 4—STATIONARY JAW (Continued)

10. Cut thread,* fitting to hex. nut as per drawing. Chamfer end of thread.
11. Lay off location of shoulder next to offset surface.
12. Set work square in shaper vise. Plane offset surface and square shoulder as per drawing.
13. Turn stock over and chamfer corner as per drawing.
14. Clamp Part No. 3 Movable Jaw and Part No. 4 Stationary Jaw together, as per assembly drawing.
15. Using Part No. 3 as a jig, spot location of center hole in Part No. 4 with $31/64''$ drill.
16. Drill center hole through Part No. 4 as per drawing. (See footnote A.)
17. With parts still clamped together, tap center hole in Part No. 4, using hole in Part No. 3 to guide tap.
18. Ream center hole in Part No. 3 as per drawing.
19. Chamfer ends of threaded hole in Part No. 4 as per drawing.
20. Place parts back together and clamp with a $1/2''$ screw through center hole.
21. Using Part No. 3 as a jig, drill the $39/64''$ holes through Part No. 4. Ream $5/8''$ through both parts.
22. Clamp Jaw Insert, Part No. 1 and Part No. 4 together, and perform drilling, tapping and chamfering operations as performed on Part No. 3.
23. File all surfaces smooth. Polish all surfaces except offset surface.*

PART No. 5—GUIDE BAR

1. Select stock as per drawing.
2. Lay off and center ends.*
3. Face ends to length as per drawing.* (See footnote D.)
4. Rough turn small diameter section.*
5. Rough turn large diameter section.
6. Finish turn section (A) of large diameter, as per drawing,* allowing .003″ for filing to diameter.
7. Finish turn remainder of large diameter section as per drawing, allowing .003″ for filing as before.
8. Finish turn small diameter section for slip fit into holes in Part No. 2—Spacer.
9. Face shoulder to length as per drawing.
10. File section A of large diameter section for press fit into Part No. 3—Movable Jaw and remainder of large diameter section for sliding fit in Part No. 4—Stationary Jaw.
11. Do not drill hole through diameter of section A until instructed.

PART No. 6—REMOVABLE BALL

1. Select stock as per drawing.
2. Place stock in lathe chuck.*
3. Face ends to length as per drawing.*
4. Center end of stock for drilling.*
5. Drill to size and depth as per drawing.*
6. Tap hole as per drawing.*
7. Cut off work to length as per drawing, plus $1/32''$ for finishing end when ball is formed.
8. Make second ball by repeating above operations No. 4 to No. 7 inclusive.

NOTE: Stop work on Part No. 6—Removable Ball. Take up and complete Part No. 7—Handle. Then use Part No. 7 as a stub arbor for finishing Part No. 6.

PART No. 7—HANDLE

1. Select stock as per drawing.
2. Place stock in lathe chuck.
3. Face ends to length as per drawing.*
4. Turn section to be threaded .010″ undersize.
5. Face shoulder to length as per drawing.
6. Arrange gearing for cutting thread as per drawing.*
7. Set lathe tool for thread cutting.*
8. Cut thread,* fitting to tapped hole in Part No. 6—Removable Ball.
9. Turn and thread other end by repeating above operations No. 4 to No. 8 inclusive.

PART No. 6—REMOVABLE BALL (Continued)

9. Screw one of the partly finished (Removable Balls) on threaded end of Part No. 7 Handle.
10. Rough turn ball with round nose tool and finish with forming tools.
11. Screw second ball on opposite end of handle. Rough and finish turn ball as before.
12. File and polish both Removable Balls and polish the handle.*

*See book "How to Run a Lathe."

SMALL BENCH VISE (Continued)

OPERATIONS:

PART No. 8—SCREW

1. Select stock as per drawing.
2. Lay off and center ends.*
3. Face ends to length as per drawing,* plus 5/16″ stock for removing the center hole in ball end. (See footnote D.)
4. Rough turn small diameter section.*
5. Rough turn large diameter section.
6. Rough turn fillet between collar and ball end.
7. Finish turn large diameter section.*
8. Finish turn ½″ diameter section, allowing .003″ for filing to diameter.
9. Turn thread section B .010″ undersize.
10. Face shoulder to length as per drawing.
11. File ½″ diameter section for running fit in ½″ reamed hole in Part No. 3—Movable Jaw.*
12. Arrange gearing for cutting thread as per drawing.
13. Set lathe tool for thread cutting.*
14. Cut thread,* fitting to tapped hole in Part No. 4—Stationary Jaw.
15. Chamfer end of thread as per drawing.
16. Place work in lathe chuck with ball end projecting.
17. Face end to length as per drawing, removing the center hole.
18. Rough turn ball with round nose tool and finish with forming tools.
19. Lay off, centerpunch and drill hole through diameter of ball. (See footnote E.)

PART No. 9—COLLAR

1. Select stock as per drawing.
2. Place stock in lathe chuck.*
3. Rough face and center stock for drilling.*
4. Drill through stock with 31/64″ drill.*
5. Ream hole as per drawing.*
6. Press ½″ arbor into work. (See footnote F.)
7. Rough face unfinished side of work.
8. Finish face sides of work to width as per drawing.
9. Turn diameter as per drawing. File and polish diameter.* File corners.

PART No. 10—WASHERS

1. Select stock as per drawing.
2. Place stock in lathe chuck.*
3. Face end* and center stock for drilling.*
4. Drill into stock ¾″ deep with 25/32″ drill.*
5. Rough and finish turn diameter and chamfer corner as per drawing.
6. Cut off washer to length as per drawing.
7. Face end of stock, chamfer corner and cut off second washer as before.

ASSEMBLING INSTRUCTIONS

1. Press the Draw Bars, Part No. 5 into the ⅝″ reamed holes in Part No. 3—Movable Jaw. (See assembly drawing.)
2. Lay off, centerpunch and drill pin holes as per drawing. Oil pins and drive same into place.
3. Assemble Part No. 3—Movable Jaw, Part No. 8—Screw and Part No. 9—Collar, as per assembly drawing.
4. Centerpunch and drill pin hole through Parts No. 8 and No. 9 as per drawing. Oil pin and drive same into place.
5. Assemble vise as per assembly drawing.

NOTE A: Drilling can be done in lathe, using drill pad in tailstock spindle and drill held in drill chuck in headstock spindle.*

NOTE B: If shaper is not available, saw slots with a hacksaw.

NOTE C: Before removing the Jaw Insert, mark both parts for convenience in assembling.

NOTE D: A relieved center would simplify the facing operation.*

NOTE E: Drilling can be done in the lathe, using crotch center in tailstock spindle and drill held in drill chuck in headstock spindle.*

NOTE F: Before pressing arbor into work, remove dirt and put oil in hole. Also clean and oil the arbor.

*See book "How to Run a Lathe."

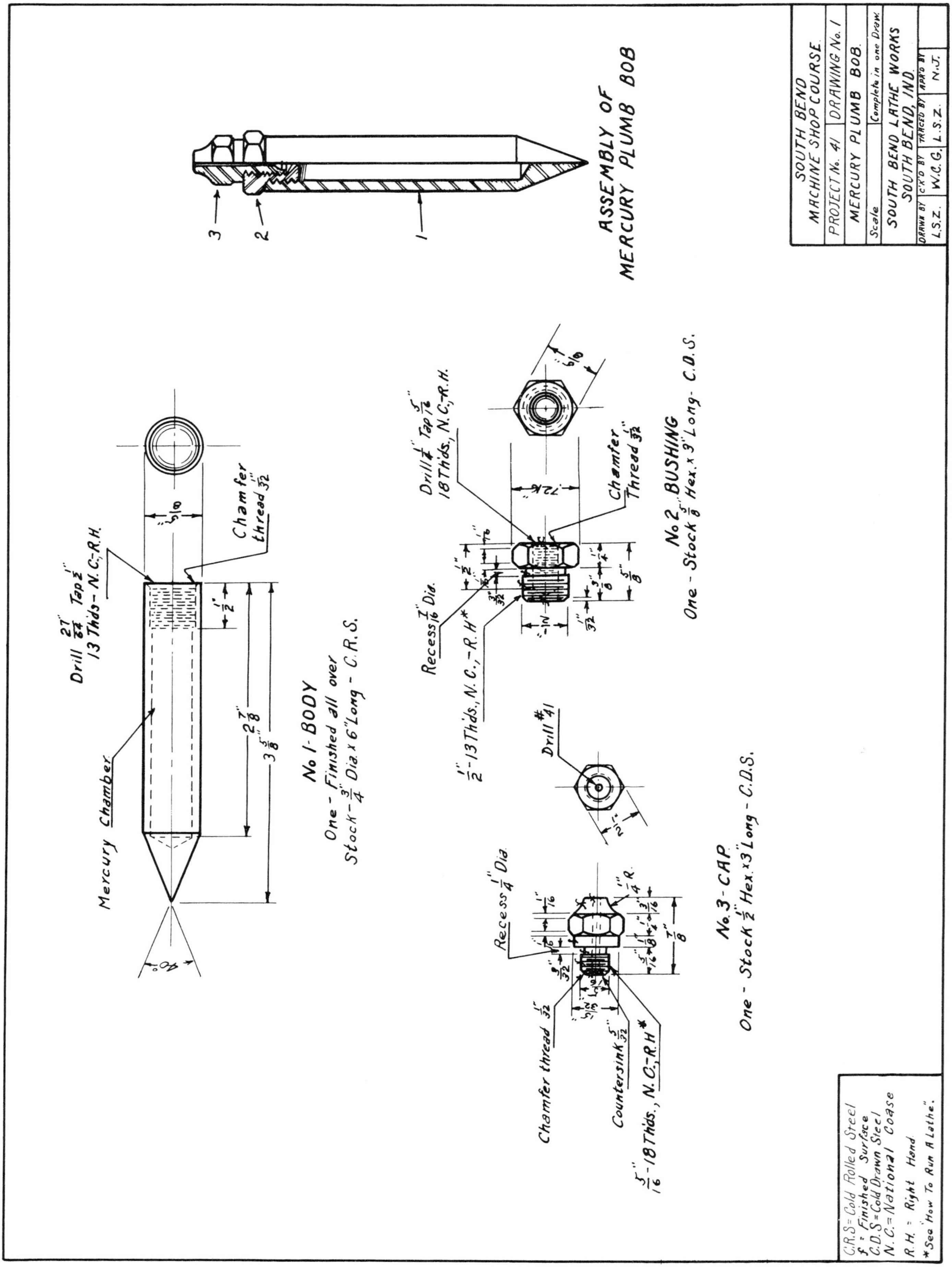
SOUTH BEND MACHINE SHOP COURSE
PROJECT No. 41
DRAWING No. 1
MERCURY PLUMB BOB
Scale
Complete in one Draw.
SOUTH BEND LATHE WORKS SOUTH BEND, IND.
DRAWN BY L.S.Z.
CH'K'D BY W.C.G.
TRACED BY L.S.Z.
APP'D BY N.J.
ASSEMBLY OF MERCURY PLUMB BOB
No 1 - BODY
One - Finished all over
Stock - 3/4" Dia. x 6" Long - C.R.S.
Drill 27/64" Tap 1/2" 13 Th'ds - N.C.-R.H.
Chamfer thread 1/32"
Mercury Chamber
No. 2 BUSHING
One - Stock 5/8" Hex. x 3" Long - C.D.S.
Drill 1/4" Tap 5/16" 18 Th'ds, N.C.-R.H.
Chamfer Thread 1/32"
Recess 7/16" Dia.
1/2" - 13 Th'ds, N.C.-R.H*
No. 3 - CAP
One - Stock 1/2" Hex. x 3" Long - C.D.S.
Drill #41
Recess 1/4" Dia
Chamfer thread 1/32"
Countersink 5/32"
5/16" - 18 Th'ds, N.C.-R.H*
C.R.S. = Cold Rolled Steel
f = Finished Surface
C.D.S. = Cold Drawn Steel
N.C. = National Coarse
R.H. = Right Hand
*See "How To Run A Lathe".

MERCURY PLUMB BOB

MATERIAL REQUIRED:

Part No. 1—Body. Cold rolled steel ¾" dia. x 6" long. One required.
Part No. 2—Bushing. Cold drawn steel ⅝" hexagon x 3" long. One required.
Part No. 3—Cap. Cold drawn steel ½" hexagon x 3" long. One required.

OPERATIONS:

PART No. 1—BODY

1. Select stock as per drawing.
2. Place stock in lathe chuck* with stock projecting 4".
3. Face end.*
4. Center end of stock for drilling.* Center drill and support end of stock with tailstock spindle.*
5. Rough and finish turn diameter as per drawing.* File and polish.*
6. Drill hole in work* with tap drill to depth as per drawing.
7. Tap end of hole as per drawing, supporting and guiding tap with tailstock center.* Chamfer end of thread.
8. Cut off work to length as per drawing, plus 1/32" stock for finishing the end.
9. Place work in lathe chuck with unfinished end projecting 1". Use strips of emery cloth to protect finished surface of work.
10. Face end to length and turn angle of point as per drawing. File and polish point.

PART No. 2—BUSHING

1. Select stock as per drawing.
2. Place stock in lathe chuck* with stock projecting 1".
3. Face end.*
4. Rough turn diameter of section to be threaded.*
5. Face shoulder and chamfer outside corners of hexagon section as per drawing.
6. Cut threading recess as per drawing.
7. Finish turn diameter of section to be threaded .010" undersize.
8. Arrange gearing for cutting thread as per drawing.*
9. Set tool for thread cutting.*
10. Cut threads.* Fit to tapped hole in Part No. 1—Body. Chamfer end of thread as per drawing.
11. Cut off work to length as per drawing, plus 1/32" stock for finishing the end.
12. Place Part No. 1—Body in lathe chuck with tapped end of work projecting ¼". Use strips of emery cloth to protect finished surface as before.
13. Screw Part No. 2—Bushing into tapped hole in Part No. 1—Body.
14. Face end of Part No. 2—Bushing and chamfer outside corners of hexagon section as per drawing.
15. Center end of work for drilling.*
16. Drill into work with tap drill to depth as per drawing.
17. Tap hole as per drawing, supporting and guiding tap with tailstock center.* Chamfer end of thread as per drawing.

PART No. 3—CAP

1. Select stock.
2. Place stock in lathe chuck with stock projecting 1⅜".
3. Face end.*
4. Rough turn both diameters, allowing 1/64" stock for finishing cut.
5. Face both shoulders* and chamfer outside corners of hexagon section as per drawing.
6. Cut threading recess as per drawing.
7. Finish turn 15/32" diameter section.* File and polish.*
8. Finish turn section to be threaded .010" undersize.
9. Center end of work for drilling.*
10. Center drill and countersink end of work as per drawing. Use a center drill with 3/32" pilot drill.
11. Drill into center of work with a No. 41 drill 1⅛" deep.*
12. Arrange gearing for cutting thread as per drawing.
13. Set lathe tool for thread cutting.*
14. Cut thread,* fitting to tapped hole in Part No. 2—Bushing. Chamfer end of thread as per drawing.*
15. Cut off work to length as per drawing plus 1/32" stock for finishing the end.
16. Screw all three parts together as per assembly drawing.
17. Place work in lathe chuck, gripping as before and allowing the hexagon parts to project out of the chuck jaws.
18. Face end of Part No. 3 to length and rough turn fillet.
19. Finish fillet with a forming tool. Chamfer outside corners of hexagon section as per drawing.
20. Round end with a file and polish fillet.*
21. Fill chamber of Part No. 1—Body with mercury and assemble Plumb Bob as per assembly drawing.

*See book "How to Run a Lathe."

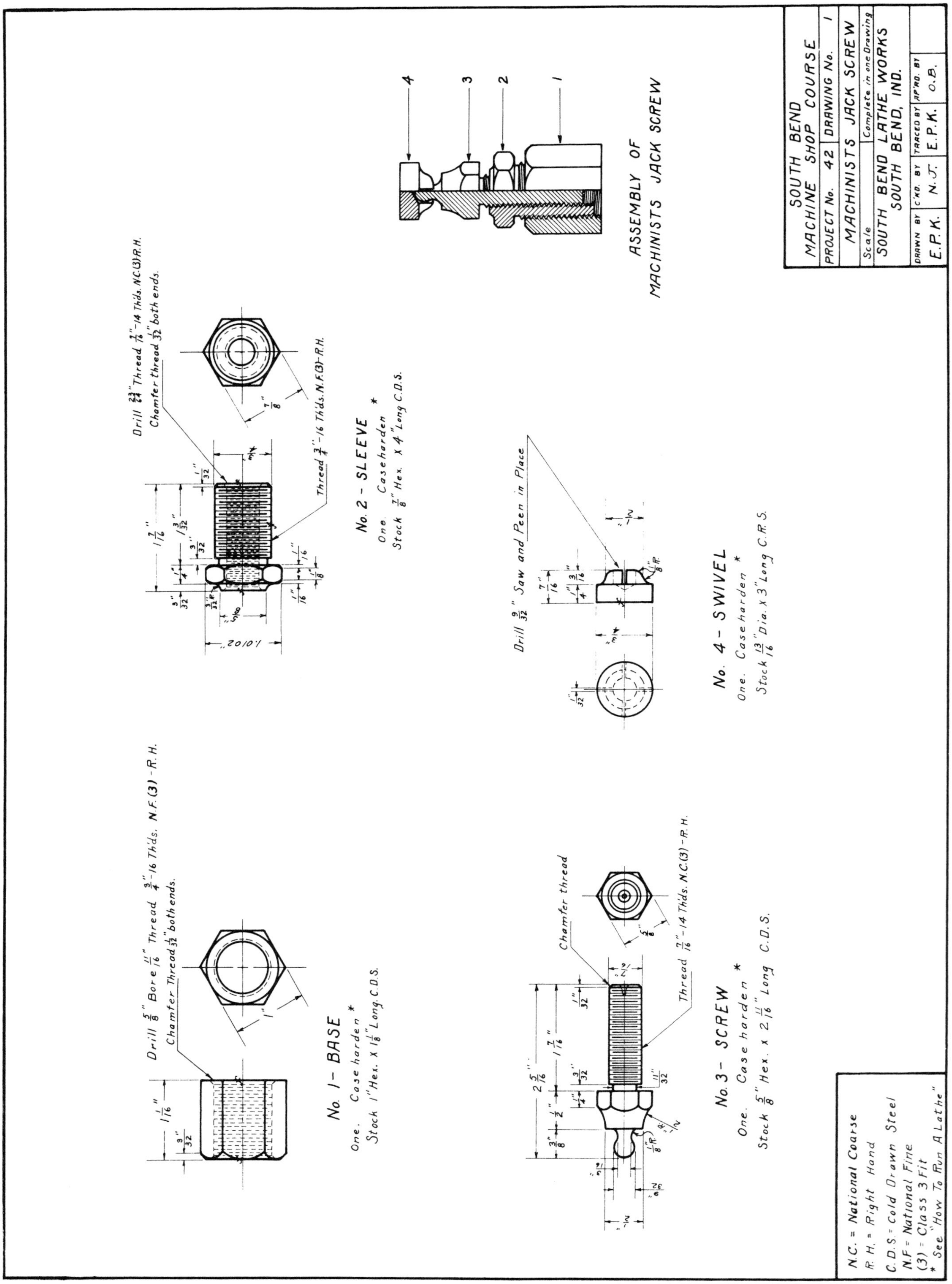
No. 1 - BASE
One. Caseharden *
Stock 1" Hex. X 1 1/8" Long. C.D.S.
Drill 5/8" Bore 11/16" Thread 3/4"-16 Th'ds. N.F. (3) - R. H.
Chamfer Thread 1/32" both ends.
No. 2 - SLEEVE
One. Caseharden *
Stock 7/8" Hex. X 4" Long C.D.S.
Drill 23/64" Thread 7/16"-14 Th'ds. N.C.(3) R.H.
Chamfer thread 1/32" both ends.
Thread 3/4"-16 Th'ds. N.F.(3)-R.H.
No. 3 - SCREW
One. Caseharden *
Stock 5/8" Hex. x 2 11/16" Long C.D.S.
Chamfer thread
Thread 7/16"-14 Th'ds. N.C.(3) - R. H.
No. 4 - SWIVEL
One. Caseharden *
Stock 13/16" Dia. x 3" Long C.R.S.
Drill 9/32" Saw and Peen in Place
ASSEMBLY OF
MACHINISTS JACK SCREW
N.C. = National Coarse
R. H. = Right Hand
C.D.S. = Cold Drawn Steel
N.F = National Fine
(3) = Class 3 Fit
* See "How To Run A Lathe"
SOUTH BEND
MACHINE SHOP COURSE
PROJECT No. 42
DRAWING No. 1
MACHINISTS JACK SCREW
Scale
Complete in one Drawing
SOUTH BEND LATHE WORKS
SOUTH BEND, IND.
DRAWN BY E.P.K.
C'KD. BY N.J.
TRACED BY E.P.K.
AP'RD. BY O.B.

MACHINIST'S JACK SCREW

MATERIAL REQUIRED:

Part No. 1—Base. Cold drawn steel 1" Hex. x 1⅛" long. One required.

Part No. 2—Sleeve. Cold drawn steel ⅞" Hex. x 4" long. One required.

Part No. 3—Screw. Cold drawn steel ⅝" Hex. x 2¹¹⁄₁₆" long. One required.

Part No. 4—Swivel. Cold rolled steel 1³⁄₁₆" dia. x 3" long. One required.

OPERATIONS:

PART No. 1—BASE

1. Select stock as per drawing.
2. Place stock in lathe chuck.*
3. Face ends to length as per drawing.*
4. Center end of stock for drilling.*
5. Drill through stock as per drawing.*
6. Bore hole as per drawing.
7. Arrange gearing for cutting thread as per drawing.*
8. Set lathe tool for internal thread cutting.*
9. Cut thread* as per drawing, allowing .005" stock for tap to remove. (See footnote A.)
10. Finish thread to size with a tap before removing work from lathe.*
11. Chamfer corners of hex. and ends of thread as per drawing.

PART No. 2—SLEEVE

1. Select stock as per drawing.
2. Place stock in lathe chuck.*
3. Face end.*
4. Rough turn diameter to be threaded.*
5. Face shoulder and cut threading recess as per drawing.
6. Center end of stock for drilling.*
7. Drill into stock as per drawing to depth of 1⁹⁄₁₆".*
8. Start thread with tap as per drawing to depth of ½".* Chamfer end of thread as per drawing.
9. Finish turn section to be threaded .010" undersize.
10. Arrange gearing for cutting thread as per drawing.*
11. Set lathe tool for thread cutting.*
12. Cut thread fitting to threaded hole in Part No. 1—Base.*
13. Chamfer outside corners of hex. section and end of thread as per drawing.
14. Cut work off to length as per drawing, plus ¹⁄₃₂" stock for finishing the end.
15. Place work in bench vise and finish tapping by hand.
16. Screw Part No. 2—Sleeve into Part No. 1—Base, with shoulder of sleeve against top of base. Place work in lathe chuck, gripping on Part No. 1—Base.
17. Face end, and rough face shoulder.*
18. Finish face shoulder, using tool ground to form fillet. File corner.
19. Chamfer outside corners of hex. section and end of thread as per drawing.

NOTE: Screw Part No. 2—Sleeve out of Part No. 1—Base, allowing the base to remain in the chuck to be used again for finishing ball end of Part No. 3—Screw.

PART No. 3—SCREW

1. Select stock as per drawing.
2. Lay off and center ends.*
3. Face ends to length as per drawing,* plus ⁵⁄₁₆" stock for removing center hole in ball end.
4. Rough turn diameter to be threaded.*
5. Face shoulder and cut threading recess as per drawing.
6. Finish turn section to be threaded .010" undersize.
7. Arrange gearing for cutting thread as per drawing.*
8. Set lathe for thread cutting.*
9. Cut thread,* fitting to tapped hole in Part No. 2—Sleeve. Chamfer end of thread as per drawing.
10. Screw Part No. 3—Screw into Part No. 2—Sleeve, with shoulder of screw against top of sleeve. Screw Part No. 2—Sleeve into Part No. 1—Base which is still in the lathe chuck.
11. Rough turn diameter of ball allowing ⁵⁄₁₆" stock for facing end to remove center hole.
12. Turn fillet next to hex. section using a forming tool.
13. Face shoulder and form fillet in one operation, using forming tool.*
14. Face end to length as per drawing, removing center hole.
15. Finish turn ball, using forming tools.

PART No. 4—SWIVEL

1. Select stock as per drawing.
2. Place stock in lathe chuck.*
3. Face end.*
4. Rough turn diameter.*
5. Center end of stock for drilling.*
6. Drill into end of stock as per drawing.*
7. Finish turn diameter as per drawing.*
8. Face shoulder and form fillet in one operation.*
9. Round end as per drawing, with a file. File corner.
10. Cut off work to length as per drawing, allowing ¹⁄₃₂" stock for finishing the end.
11. Place work in lathe chuck and face unfinished end to length as per drawing. File corner.
12. Place work in bench vise and saw slots as per drawing.
13. Caseharden the four parts all over.*
14. Assemble the jack and peen swivel as per assembly drawing.

NOTE A: If tap is not available cut thread to full depth.

*See book "How to Run a Lathe."

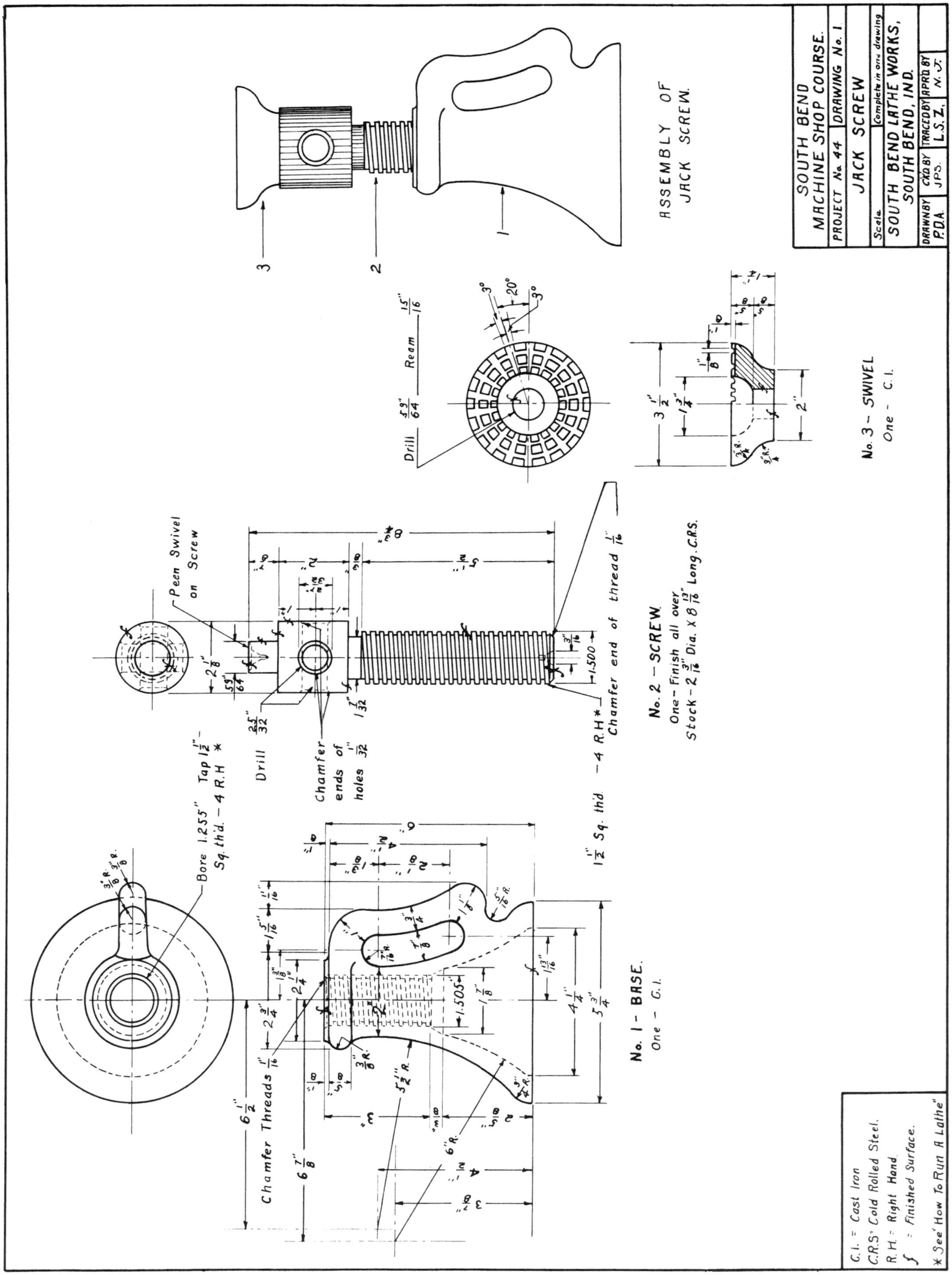
SOUTH BEND MACHINE SHOP COURSE.
PROJECT No. 44 DRAWING No. 1
JACK SCREW
Scale
Complete in one drawing
SOUTH BEND LATHE WORKS, SOUTH BEND, IND.
DRAWN BY P.D.A.
CK'D BY JPS.
TRACED BY L.S.Z.
APP'D BY N.J.
ASSEMBLY OF JACK SCREW
Drill 59/64" Ream 15/16"
No. 3 - SWIVEL
One - C.I.
Peen Swivel on Screw
Drill 25/32
Chamfer ends of 1" holes 1/32
Chamfer end of thread 1/16"
1 1/2" Sq. thd. - 4 R.H.*
No. 2 - SCREW
One - Finish all over
Stock - 2 3/16" Dia. X 8 13/16" Long. C.R.S.
Bore 1.255" Tap 1 1/2" Sq. th'd. - 4 R.H *
Chamfer Threads 1/16"
No. 1 - BASE.
One - C.I.
C.I. = Cast Iron
C.R.S. = Cold Rolled Steel.
R.H. = Right Hand.
f = Finished Surface.
* See "How To Run A Lathe"

JACK SCREW

MATERIAL REQUIRED:

Part No. 1—Base. Cast iron. One required.

Part No. 2—Screw. Cold rolled steel 2$\frac{3}{16}$″ dia. x 8$\frac{3}{16}$″ long. One required.

Part No. 3—Swivel. Cast iron. One required.

OPERATIONS:

PART No. 1—BASE

1. Remove sand and rough projections from casting.
2. Turn casting end for end in chuck and face small end.
3. Place casting in lathe chuck and face the large end to length.
4. Bore cored hole to size.
5. Cut thread,* allowing .005″ stock for tap to remove.
6. Finish thread to size with a tap. Chamfer end of thread. Support and guide tap with tailstock center.

PART No. 2—SCREW

1. Select stock.
2. Lay off and center ends.*
3. Face ends to length.*
4. Rough turn all diameters allowing $\frac{1}{64}$″ stock for finishing cut.
5. Face shoulders to length and cut threading recess.*
6. Finish turn and file all diameters to size. File corners and polish large diameter section.
7. Cut thread,* fitting to tapped hole in Part No. 1—Base. Chamfer end of thread.
8. Lay off and center punch location of holes through large diameter section.
9. Drill holes through large diameter section. Chamfer ends of holes.

PART No. 3—SWIVEL

1. Remove sand and rough projections from casting.
2. Place casting in lathe chuck and face small end of casting to length.
3. Center casting for drilling and drill hole through same.
4. Ream hole to size.
5. Assemble jack as per assembly drawing.

*See book "How to Run a Lathe."

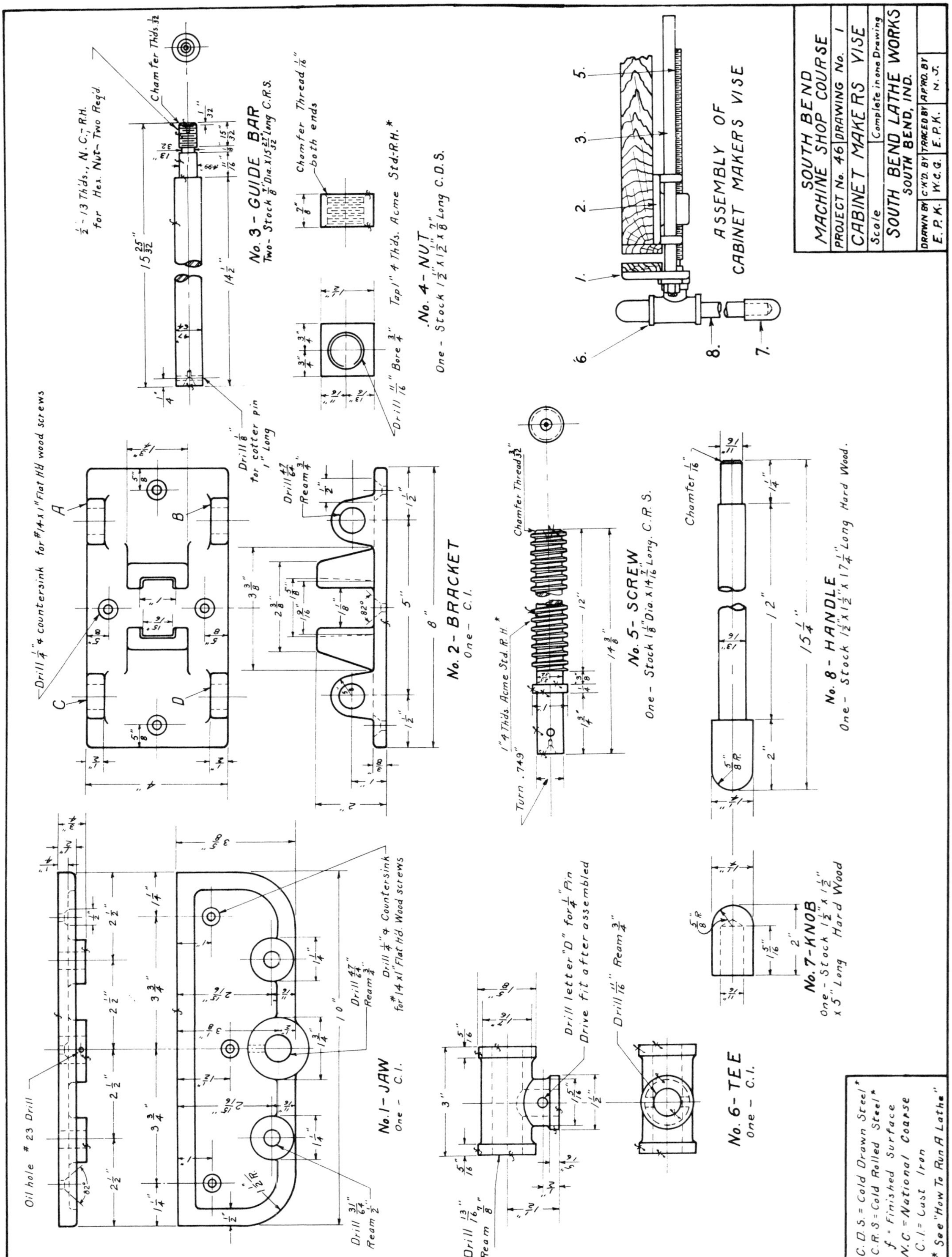
SOUTH BEND MACHINE SHOP COURSE
PROJECT No. 46
DRAWING No. 1
CABINET MAKERS VISE
Scale
Complete in one Drawing
SOUTH BEND LATHE WORKS
SOUTH BEND, IND.
DRAWN BY E.P.K.
CH'K'D. BY W.C.G.
TRACED BY E.P.K.
AP'RD. BY N.J.
ASSEMBLY OF CABINET MAKERS VISE
No. 3 - GUIDE BAR
No. 4 - NUT
No. 2 - BRACKET One - C.I.
No. 5 - SCREW
No. 8 - HANDLE
No. 1 - JAW One - C.I.
No. 6 - TEE One - C.I.
No. 7 - KNOB
C.D.S. = Cold Drawn Steel
C.R.S. = Cold Rolled Steel
f = Finished Surface
N.C = National Coarse
C.I. = Cast Iron
* See "How To Run A Lathe"

CABINETMAKER'S VISE

MATERIAL REQUIRED:

Part No. 1—Jaw. Cast iron. One required.

Part No. 2—Bracket. Cast iron. One required.

Part No. 3—Guide Bar. Cold rolled steel 7/8" dia. x 15 27/32" long. Two required.

Part No. 4—Nut. Cold drawn steel 1 1/2" sq. x 15/16" long. One required.

Part No. 5—Screw. Cold rolled steel 1 1/8" dia. x 14 7/16" long. One required.

Part No. 6—Tee. Cast iron. One required.

Part No. 7—Knob. Hardwood 1 1/2" sq. x 5" long. One required.

Part No. 8—Handle. Hardwood 1 1/2" sq. x 17 1/4" long. One required.

Wood Screws size No. 14 x 1" long. Seven required.

Cotter Pins 1/8" dia. x 1" long. Two required.

OPERATIONS:

PART No. 1—JAW

1. Remove sand and rough projections from casting.
2. Place casting in shaper vise and plane top edge.
3. Set casting level and plane flat surface.
4. Turn casting over and plane bosses on irregular side.
5. Lay off and center punch location of holes to be drilled.
6. Drill and ream the 3/4" and 1/2" holes. Drill and countersink 1/4" holes for wood screws. Drill oil hole.

PART No. 2—BRACKET

1. Remove sand and rough projections from casting.
2. Set casting lever in shaper vise and plane flat surface.
3. Lay off and center punch location of holes to be drilled.
4. Drill and countersink 1/4" holes for wood screws.
5. Drill and ream holes in lugs A.B. and C.D. between centers in lathe.*

PART No. 3—GUIDE BAR

1. Select two pieces of stock. Perform following operations on both parts.
2. Lay off and center ends.*
3. Face ends to length.*
4. Rough turn small diameter section.
5. Rough turn large diameter section.
6. Finish turn large diameter section, allowing .003" for filing to diameter. File and polish to size.
7. Face shoulder to length.
8. Cut threading recess.
9. Finish turn 1/2" diameter section, allowing .003" for filing to size. File for slip fit into 1/2" reamed holes in Part No. 1—Jaw.
10. Finish turn section to be threaded .010" undersize.
11. Cut thread to fit a 1/2" hexagon nut.*
12. Chamfer end of thread.
13. Drill cotter pin hole through large diameter.

*See book "How to Run a Lathe."

CABINETMAKER'S VISE (Continued)

OPERATIONS:

PART No. 4—NUT

1. Select stock.
2. Lay off and center punch center location of hole.
3. Place stock in lathe chuck, truing to center punch mark.*
4. Face side of stock.
5. Center stock for drilling.*
6. Drill hole through stock.*
7. Bore hole to size.
8. Cut thread, allowing .005″ stock for tap to remove.*
9. Finish thread to size with a tap. Support and guide tap with tailstock spindle.*
10. Chamfer end of thread.
11. Turn stock around in chuck and face other side.
12. Chamfer end of thread as before.
13. Remove burrs and sharp corners with a file.

PART No. 5—SCREW

1. Select stock.
2. Lay off and center ends.*
3. Face ends to length.*
4. Rough turn section to be threaded.
5. Rough turn small diameter section.
6. Face shoulder to length.
7. Cut recess between collar and section to be threaded.
8. Finish turn small diameter section, allowing .003″ for filing to size.
9. File and polish small diameter section for running fit in ¾″ reamed hole in Part No. 1—Jaw.*
10. Finish turn section to be threaded.
11. Cut thread to fit tapered hole in Part No. 4—Nut.*
12. Chamfer end of thread.

PART No. 6—TEE

1. Remove sand and rough projections from casting.
2. Place casting in lathe chuck in position to bore the long hole.*
3. Face end of casting.*
4. Chamfer end of cored hole for starting drill.*
5. Drill through cored hole with 3 or 4 lip drill.*
6. Bore hole allowing .005″ to .010″ for reaming to size.
7. Ream hole to size.*
8. Press a mandrel in reamed hole in casting.*
9. Face unfinished end of casting to length.
10. Rough and finish turn 1⅝″ diameters. File corners. File and polish diameters.
11. Remove mandrel and place casting in lathe chuck in position to bore short hole.
12. Face end to length and chamfer end of cored hole for starting drill.*
13. Drill short cored hole to meet the long one. Use 3 or 4 lip drill.*
14. Rough and finish turn the 1½″ diameter.
15. Bore hole allowing .005″ to .010″ for reaming to size.
16. Ream hole to size.*

*See book "How to Run a Lathe."

CABINETMAKER'S VISE (Continued)

OPERATIONS:

PART No. 7—KNOB

1. Select stock.
2. Place stock in lathe chuck.*
3. Face and center end of stock for drilling.*
4. Drill hole in end of stock to depth.*
5. Turn diameter and sand surface smooth.
6. Cut off work to length, plus ⅛″ stock for finishing end.
7. Chuck another piece of hardwood and turn end to fit in Part No. 7—Knob. This should be a light drive fit.
8. Place knob on stud arbor and face end to length.
9. Round end and sand surface smooth.

PART No. 8—HANDLE

1. Select stock.
2. Lay off centers on ends of stock.*
3. Place spur center in headstock spindle, and cup center in tailstock spindle.
4. Place stock between the centers. The stock is driven by the spur center.
5. Turn large diameter section.
6. Turn center section allowing 1″ on length of large diameter section for removing the center hole.
7. Turn small diameter section for slip fit in hole in Part No. 7—Knob.
8. Face shoulders to length and sand surface smooth.
9. Saw off both ends allowing ⅛″ on each end for finishing.
10. Place handle in lathe chuck with large end projecting.
11. Face end to length. Round end and sand surface smooth.
12. Turn work end for end and face small end to length. Chamfer corner.

ASSEMBLING INSTRUCTIONS

1. Assemble Part Nos. 1, 5, and 6 as per assembly drawing.
2. Drill and pin Part Nos. 5 and 6. Oil pin before driving same into hole.
3. Assemble Part Nos. 6, 7, and 8. Glue Parts No. 7 and No. 8 together.
4. Attach vise to bench as per assembly drawing.

*See book "How to Run a Lathe."

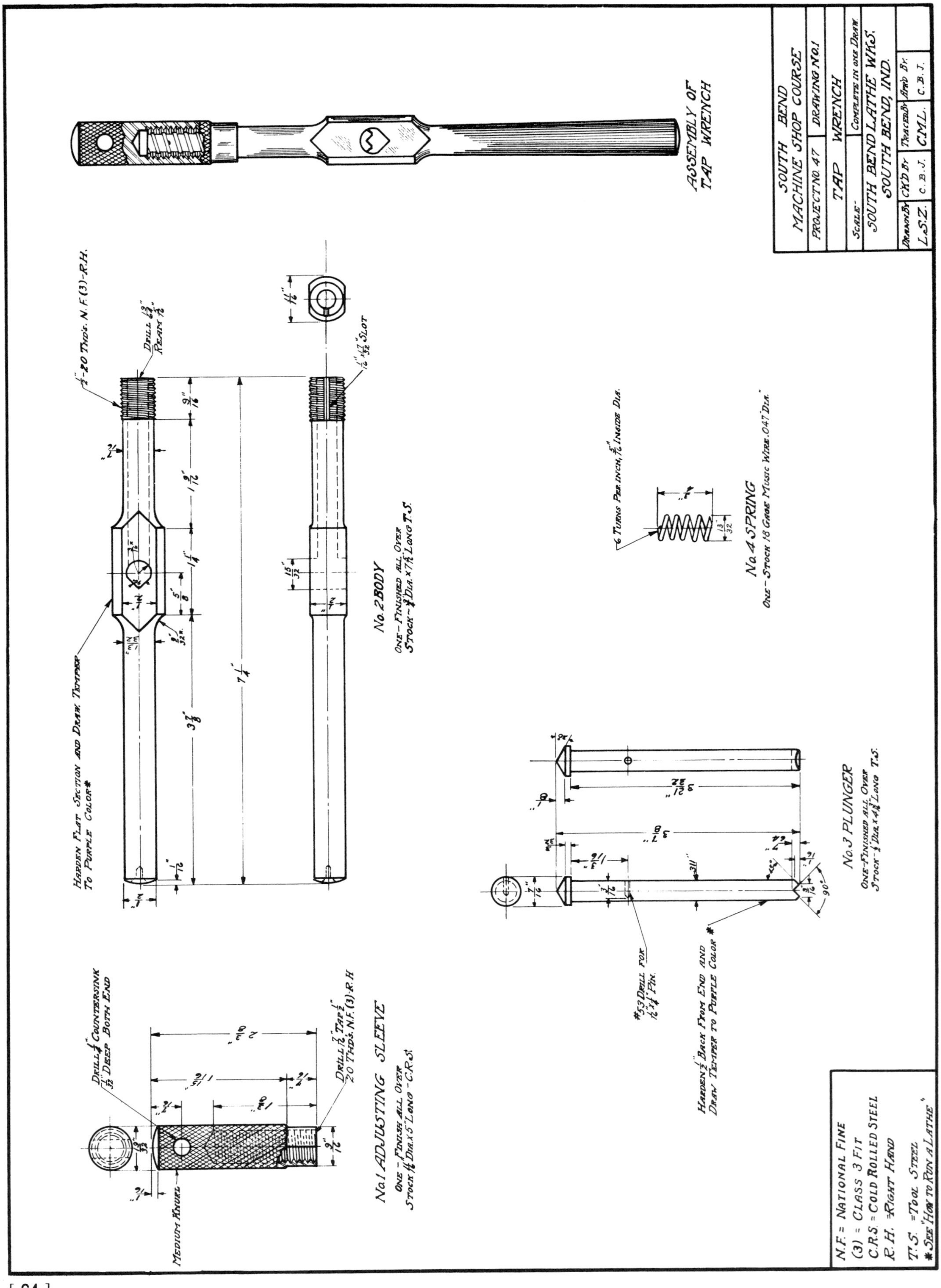
SOUTH BEND
MACHINE SHOP COURSE
PROJECT NO. 47
DRAWING NO. 1
TAP WRENCH
SCALE-
COMPLETE IN ONE DRAW.
SOUTH BEND LATHE WKS.
SOUTH BEND, IND.
DRAWN BY
L.S.Z.
CK'D BY
C.B.J.
TRACED BY
CML.
APP'D BY
C.B.J.
ASSEMBLY OF
TAP WRENCH
HARDEN FLAT SECTION AND DRAW TEMPER
TO PURPLE COLOR *
No. 2 BODY
ONE - FINISHED ALL OVER
6 TURNS PER INCH, 5/16" INSIDE DIA.
No. 4 SPRING
ONE - STOCK 18 GAGE MUSIC WIRE .047" DIA.
No. 3 PLUNGER
#53 DRILL FOR PIN.
HARDEN BACK FROM END AND
DRAW TEMPER TO PURPLE COLOR *
No. 1 ADJUSTING SLEEVE
ONE - FINISH ALL OVER
MEDIUM KNURL
N.F. = NATIONAL FINE
(3) = CLASS 3 FIT
C.R.S. = COLD ROLLED STEEL
R.H. = RIGHT HAND
T.S. = TOOL STEEL
* SEE "HOW TO RUN A LATHE"

ADJUSTABLE TAP WRENCH

MATERIAL REQUIRED:

Part No. 1—Adjusting Sleeve. Cold rolled steel 11/16″ dia. x 5″ long. One required.
Part No. 2—Body. Tool steel 3/4″ dia. x 75/16″ long. One required.
Part No. 3—Plunger. Tool steel 1/2″ dia. x 43/16″ long. One required.
Part No. 4—Spring. Music wire (18 gauge .047″ dia.) 2 1/8″ required for each spring.

OPERATIONS:

PART No. 1—ADJUSTING SLEEVE

1. Select stock.
2. Place stock in lathe chuck.*
3. Face end.* Center drill and adjust tailstock to support end of work.
4. Rough turn diameters,* allowing 1/64″ stock for finishing cut.
5. Finish turn and knurl large diameter section.*
6. Finish turn small diameter section. File corner. File and polish small diameter section.*
7. Center end of work for drilling.*
8. Drill into end of work with tap drill to depth.*
9. Tap hole to size, supporting and guiding tap with tailstock center.*
10. Cut off work to length, plus 1/32″ stock for finishing the end.
11. Place work in lathe chuck,* using sheet brass strips to protect knurled surface. Face and round end to length. File and polish rounded end.*
12. Lay off, centerpunch, drill and countersink hole through diameter.

PART No. 2—BODY

1. Select stock.
2. Lay off and center ends.*
3. Face ends to length.*
4. Rough turn all diameters,* allowing 1/64″ stock for finishing cut. Rough turn fillets at shoulders.*
5. Finish turn all diameters and form fillets at shoulders.*
6. Cut thread,* fitting to tapped hole in Part No. 1—Adjusting Sleeve.
7. Plane or mill flat surfaces of center section.
8. Lay off, centerpunch and drill hole through center of flat section.
9. Arrange lathe to support work with center rest, on 7/16″ diameter section.*
10. Center end of work for drilling.*
11. Drill and ream hole through handle to meet drilled hole through flat section.*
12. Saw 1/16″ slot through threaded section.
13. File 90° notch in side of hole through flat section. File and polish all machined surfaces.
14. Harden flat section and draw temper.* Re-polish all machined surfaces.

PART No. 3—PLUNGER

1. Select stock.
2. Lay off and center ends.*
3. Face ends to length,* plus 5/8″ stock for removing the center holes.
4. Rough turn diameters,* allowing 1/64″ stock for finishing cut, and 5/16″ on each end for removing the center holes.
5. Face shoulder to length.
6. Finish turn diameters,* allowing .003″ stock on small diameter for filing to size. File for slip fit in reamed hole through handle of Part No. 2—Body.
7. Place work in lathe chuck* with small end projecting and face end to length, removing the center hole.
8. Turn work end for end and face large end to length, removing center hole as before.
9. Turn 30° angle on large diameter section.*
10. File 90° notch and 45° angles on end of small diameter section.
11. Lay off, centerpunch and drill hole for 1/16″ pin.
12. Harden 1/2″ back from small end and draw temper.*

PART No. 4—SPRING

1. Select a 1/4″ cold rolled steel rod 4″ long and drill a hole through diameter 1 1/2″ from end with a No. 66 drill.
2. Place rod in lathe chuck with drilled end projecting 1 3/4″.
3. Center drill and support end of rod with tailstock center.*
4. Run music wire (18 gauge) between two wood fibre blocks clamped in tool post. Bend a hook on end of wire 1/4″ long and at right angles.
5. Place hook through hole in rod (from bottom side) and run cross slide back until wire is stretched tight.
6. Throw in back gears and shift belt to large step of spindle cone.*
7. Arrange lathe to cut 6 threads per inch, left hand.*

   NOTE: The wire should be stretched at right angles to the rod when winding of spring begins. When spring is wound, back lathe spindle up by hand until tension is relieved, before cutting the wire.
8. Wind spring the same as cutting a left hand thread.
9. Cut wire close to hook and remove spring from rod. Cut spring to length and grind ends.
10. Harden in oil. Burn oil off three times to draw temper.

*See book "How to Run a Lathe."

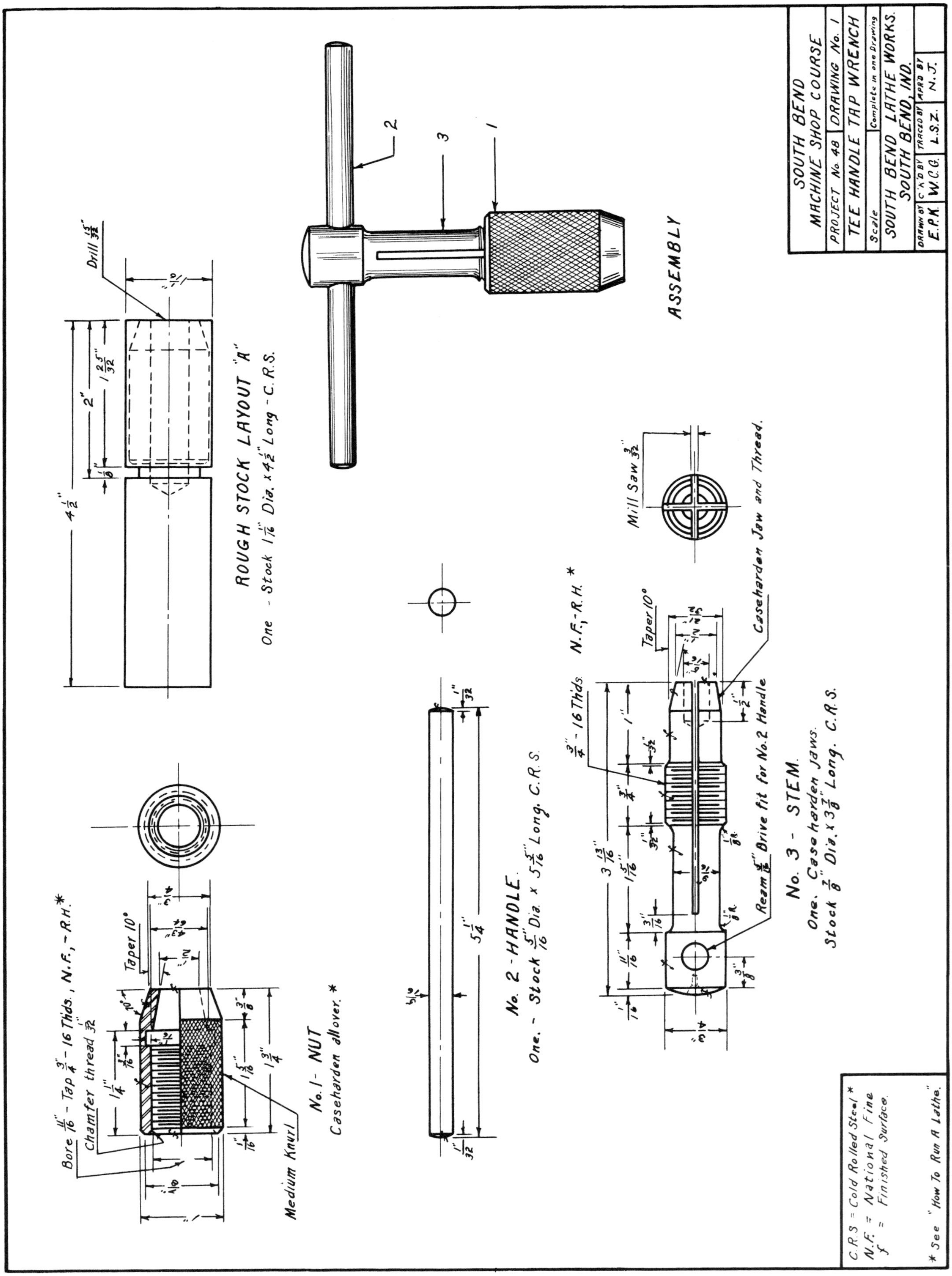
SOUTH BEND
MACHINE SHOP COURSE
PROJECT No. 48
DRAWING No. 1
TEE HANDLE TAP WRENCH
Scale
Complete in one Drawing
SOUTH BEND LATHE WORKS.
SOUTH BEND, IND.
DRAWN BY E.P.K.
CH'K'D BY W.C.G.
TRACED BY L.S.Z.
APPR'D BY N.J.
ASSEMBLY
ROUGH STOCK LAYOUT "A"
One - Stock 1 1/16" Dia. x 4 1/2" Long - C.R.S.
Drill 15/32"
No. 1 - NUT
Caseharden allover. *
Bore 11/16" - Tap 3/4" - 16 Th'ds., N.F., - R.H. *
Chamfer thread 1/32"
Taper 10°
Medium Knurl
No. 2 - HANDLE
One. - Stock 5/16" Dia. x 5 5/16" Long. C.R.S.
No. 3 - STEM.
One. Caseharden Jaws.
Stock 7/8" Dia. x 3 7/8" Long. C.R.S.
3/4" - 16 Th'ds N.F., - R.H. *
Caseharden Jaw and Thread.
Mill Saw 3/32"
Ream 5/16" Drive fit for No. 2 Handle
C.R.S. = Cold Rolled Steel *
N.F. = National Fine
f = Finished Surface.
* See "How To Run A Lathe."

"T" HANDLE TAP WRENCH

MATERIAL REQUIRED:

Part No. 1—Nut. Cold rolled steel 1$\frac{1}{16}$" dia. x 4½" long. One required.

Part No. 2—Handle. Cold rolled steel $\frac{5}{16}$" dia. x 5$\frac{5}{16}$" long. One required.

Part No. 3—Stem. Cold rolled steel ⅞" dia. x 3⅞" long. One required.

OPERATIONS:

PART No. 1—NUT

1. Select stock as per layout A.
2. Place stock in lathe chuck.*
3. Face end.*
4. Center drill and support end of work with tailstock center.*
5. Rough and finish turn diameter. Rough turn taper.
6. Knurl straight section.*
7. Finish turn taper section. File corner.
8. Center end of work for drilling.*
9. Drill into end of work as per stock layout A.*
10. File tapered section. Polish tapered section and face of work.*
11. Cut off work to length as per stock layout A.
12. Place work in lathe chuck*, with unfinished end projecting.
13. Face unfinished end to length.
14. Drill 1⅛" deep with ⅝" drill.
15. Bore hole to size and depth.
16. Bore threading recess.
17. Bore taper hole.
18. Cut thread allowing .005" stock for tap to remove.*
19. Finish thread to size with a tap. Support and guide tap with tailstock center.*
20. Chamfer end of thread and outside corner. Polish chamfered corner and face of work.*
21. Caseharden all over.*

PART No. 2—HANDLE

1. Select stock.
2. Place stock in lathe chuck.*
3. Face and round ends to length.*

PART No. 3—STEM

1. Select stock.
2. Lay off and center ends.*
3. Face ends to length.*
4. Rough turn all diameters, allowing $\frac{1}{64}$" stock for finishing cut.
5. Round end of head section.
6. Finish turn all diameters except the taper section. File corners.
7. Turn section to be threaded .010" undersize.
8. Cut thread, fitting to tapped hole in Part No. 1—Nut. Chamfer ends of thread.
9. Finish turn taper section, fitting to taper hole in Part No. 1—Nut.
10. Drill hole in jaw end of work between centers in lathe.*
11. Lay off, center punch and drill and ream hole through diameter of head section.*
12. Saw slots, using milling machine.
13. Caseharden jaws and thread.*
14. Assemble wrench as per assembly drawing. Polish handle after wrench is assembled.*

*See book "How to Run a Lathe."

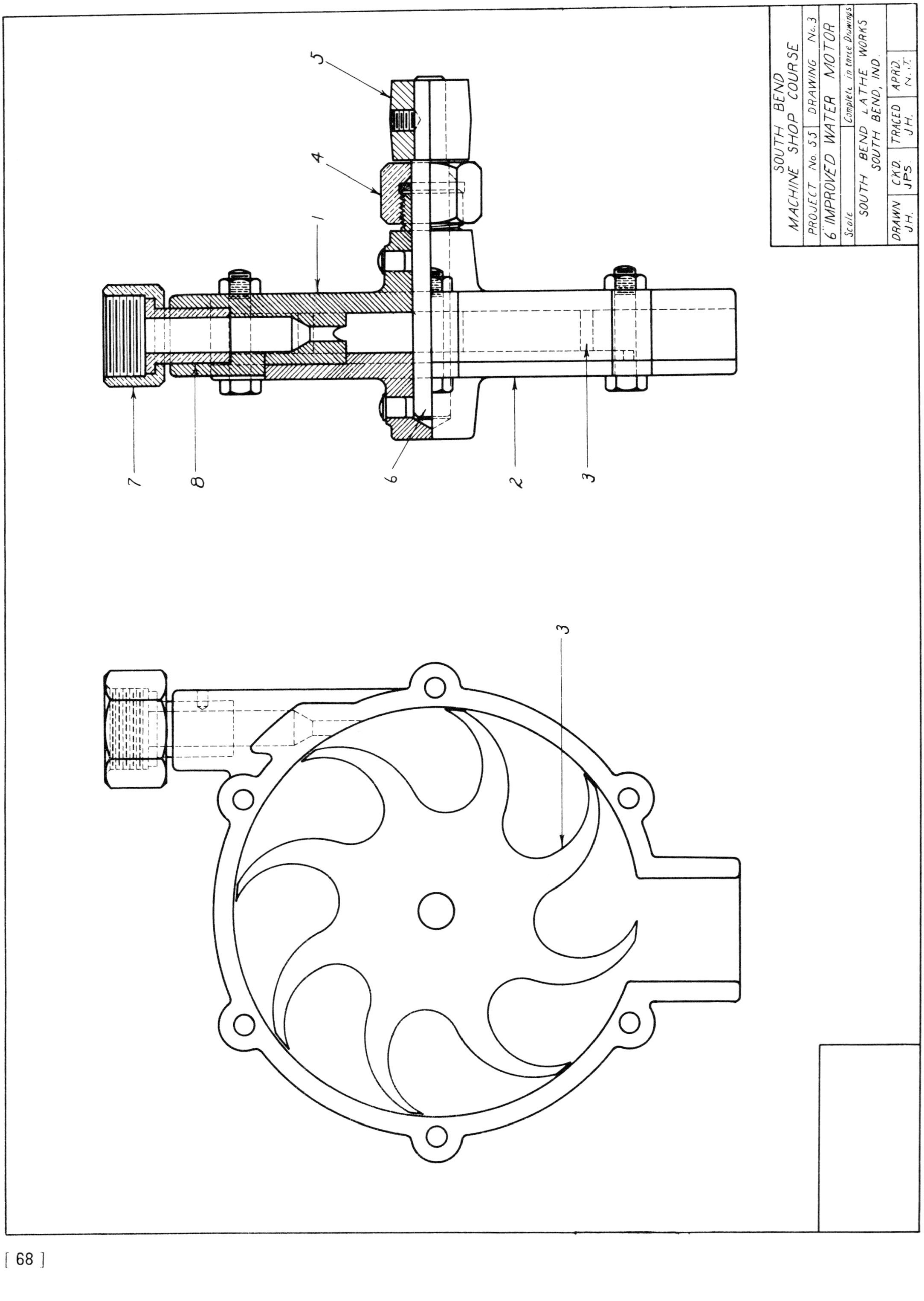
SOUTH BEND
MACHINE SHOP COURSE
PROJECT No. 55
DRAWING No. 3
6" IMPROVED WATER MOTOR
Scale
Complete in three Drawings
SOUTH BEND LATHE WORKS
SOUTH BEND, IND.
DRAWN J.H.
CK'D. JPS.
TRACED J.H.
APRD. N.T.
1
2
3
4
5
6
7
8
3

6″ IMPROVED WATER MOTOR

MATERIAL REQUIRED:

Part No. 1—Wheel Housing. Cast iron. One required.
Part No. 2—Housing Cover. Cast iron. One required.
Part No. 3—Water Wheel. Cast brass. One required.
Part No. 4—Stuffing box. Brass 1⅟16″ hexagon x 13⁄16″ long. One required.
Part No. 5—Pulley. Cast iron. One required.
Part No. 6—Spindle. Machinery steel 9⁄16″ diameter x 4¼″ long. One required.
Part No. 7—Coupling Nut. Brass 1¼″ hexagon x 13⁄16″ long. One required.
Part No. 8—Coupling Nipple. Brass 1″ diameter x 1⅛″ long. One required.
Part No. 9—Cold Rolled Steel Pin ⅛″ dia. x ¼″ long.
Headless Set Screw ¼″ diameter x ¼″ long. National Coarse—R.H. One required.
Hexagon Head Cap Screws ¼″ diameter x 1¼″—28 threads. National Fine—R.H. Six required.
Hexagon Nuts ¼″—28 threads. National Fine—R.H. Six required.

OPERATIONS:

PART No. 1—WHEEL HOUSING

1. Remove sand and rough projections from casting.
2. Place casting in lathe chuck* with open side out.
3. Rough bore inside of casting and rough face both surfaces allowing 1⁄64″ stock for finishing cut.
4. Center casting for drilling* and drill hole through bearing boss.
5. Finish face both surfaces to dimensions, and finish bore inside diameter to size.
6. Bore bearing hole, allowing .005″ to .010″ for reaming to size.
7. Machine ream bearing hole to standard size.*
8. Press a mandrel in casting and face end of bearing boss to length.
9. Rough turn diameter of section to be threaded.
10. Face shoulder to length and finish turn section to be threaded .010″ undersize.

    NOTE: Do not cut thread until instructed.
11. Lay off and center punch locations of all holes to be drilled.
12. Drill hole in bearing boss for No. O-A Ball Oiler.
13. Drill holes for screws to hold the housing cover in place. Spot face for ¼″ nuts.
14. Chamfer outside end of bearing hole with a 45 degree countersink having 7⁄16″ pilot.
15. Drill 3⁄16″ hole for water intake. Redrill with 7⁄16″ drill to depth.
16. Redrill with 43⁄64″ drill to depth, and square shoulder with a flat drill. Ream hole to size.

    NOTE: Do not drill ⅛″ pin hole until after motor is assembled.

PART No. 2—HOUSING COVER

1. Remove sand and rough projections from casting.
2. Place casting in lathe chuck* with straight side out.
3. Rough face surface and center casting for drilling.*
4. Drill hole into bearing boss to depth.*
5. Rough face shoulder and rough turn the offset diameter.
6. Finish face surface and finish face shoulder to height.
7. Finish turn the offset diameter to size, fitting to inside diameter of Part No. 1—Wheel Housing.

*See book "How to Run a Lathe."

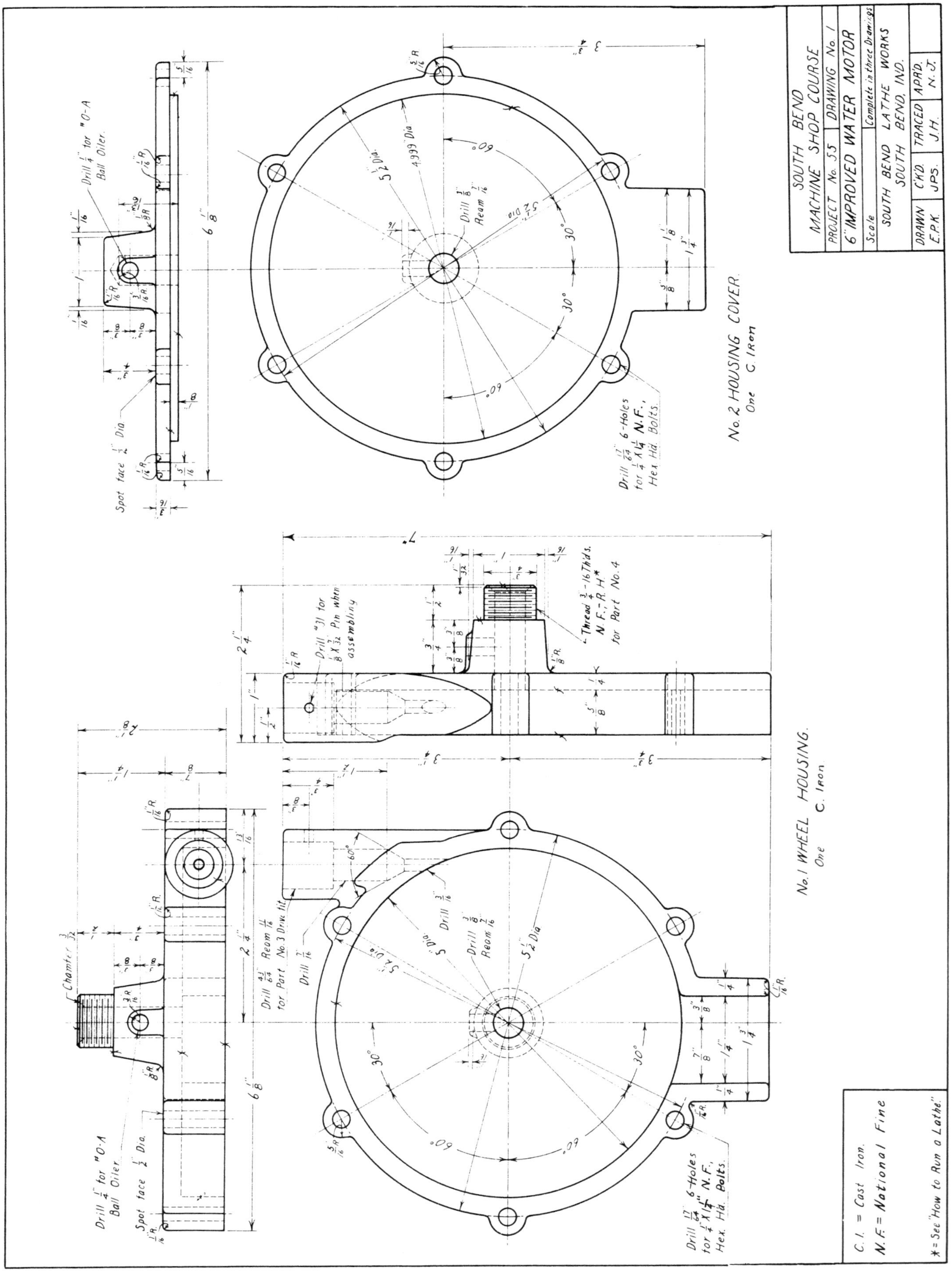
SOUTH BEND MACHINE SHOP COURSE
PROJECT No. 55
DRAWING No. 1
6" IMPROVED WATER MOTOR
Complete in three Drawings
Scale
SOUTH BEND LATHE WORKS SOUTH BEND, IND.
DRAWN E.P.K.
C'K'D. JPS.
TRACED J.H.
APP'D. N.J.
No. 2 HOUSING COVER
One C. Iron
No. 1 WHEEL HOUSING
One C. Iron
C. I. = Cast Iron.
N. F. = National Fine
* = See "How to Run a Lathe."

6″ IMPROVED WATER MOTOR (Continued)

OPERATIONS:

PART No. 2—HOUSING COVER (Continued)

8. Bore bearing hole, allowing .005″ to .010″ stock for reaming to size.
9. Machine ream bearing hole to standard size.*
10. Clamp Part No. 1—Wheel Housing to Part No. 2—Housing Cover as per assembly drawing.
11. Using Part No. 1 as a jig, drill screw holes in Part No. 2 for holding the cover in place. Spot face for heads of cap screws.
12. Drill hole in bearing boss for No. O-A Ball Oiler.

PART No. 3—WATER WHEEL

1. Remove sand and rough projections from casting.
2. Lay off and center punch location of hole in center of wheel.
3. Clamp casting on large face plate.* Use a center test indicator and adjust work so center punch mark runs true.
4. Center casting for drilling and drill hole through center.*
5. Bore hole allowing .005″ to .010″ stock for reaming to size.
6. Machine ream hole to standard size.*
7. Press a mandrel in casting. Rough face both sides of work and rough turn diameter, allowing 1/64″ stock for finishing cut. (See footnote A.)
8. Use right and left-hand tools and finish face both sides of work to width without changing position of arbor between centers.
9. Finish turn diameter to size.

PART No. 4—STUFFING BOX

1. Select stock.
2. Place stock in lathe chuck.
3. Face end* and chamfer outside corners.
4. Center casting for drill and drill hole through stock.*
5. Bore inside diameter of section to be threaded and bore threading recess.
6. Chamfer corner of inside shoulder.
7. Cut thread* allowing .005″ stock for tap to remove.
8. Finish thread to size with a tap, supporting and guiding tap with tailstock center. Chamfer end of thread.
9. Turn work end for end in chuck and face unfinished end to length. Chamfer outside corners.
10. Place Part No. 1—Wheel Housing back in lathe and cut thread on bearing boss, fitting to Part No. 4—Stuffing Box.

PART No. 5—PULLEY

1. Remove sand and rough projections from casting.
2. Place casting in lathe chuck.*
3. Face and center casting for drilling.*
4. Drill and ream hole through casting to size.*
5. Press a mandrel in casting and face unfinished end to length.
6. Rough turn crown on pulley face allowing 1/64″ stock for finishing cut.
7. Finish turn crown to size. File corners. File and polish crowned face of pulley.*
8. Center punch, drill and tap set screw hole.

*See book "How to Run a Lathe."

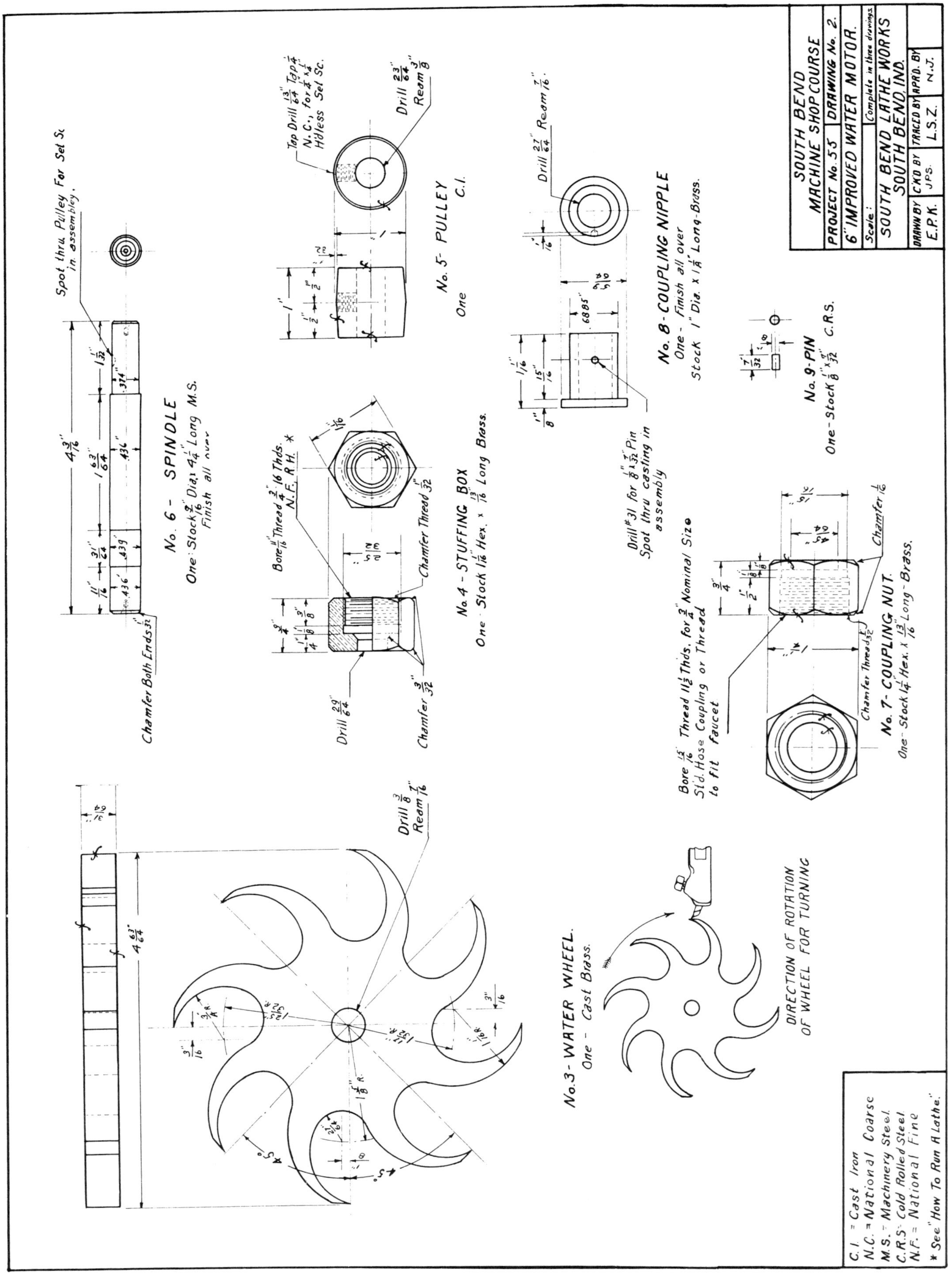
SOUTH BEND MACHINE SHOP COURSE
PROJECT No. 55
DRAWING No. 2.
6" IMPROVED WATER MOTOR.
Scale:
Complete in three drawings.
SOUTH BEND LATHE WORKS SOUTH BEND, IND.
DRAWN BY E.P.K.
C'K'D BY JPS.
TRACED BY L.S.Z.
APRD. BY N.J.
No. 6 - SPINDLE
One: Stock 3/8" Dia. x 4 1/4" Long M.S.
Chamfer Both Ends 1/32"
Spot thru Pulley For Set Sc. in assembley.
No. 5 - PULLEY C.I.
One
No. 4 - STUFFING BOX
One Stock 1 1/16" Hex. x 13/16" Long Brass.
No. 8 - COUPLING NIPPLE
One - Finish all over
Stock 1" Dia. x 1 1/4" Long-Brass.
No. 9 - PIN
One - Stock 1/8" x 9/32" C.R.S.
No. 7 - COUPLING NUT.
One - Stock 1 1/4" Hex. x 13/16" Long - Brass.
No. 3 - WATER WHEEL.
One - Cast Brass.
DIRECTION OF ROTATION OF WHEEL FOR TURNING
C.I. = Cast Iron
N.C. = National Coarse
M.S. = Machinery Steel
C.R.S. = Cold Rolled Steel
N.F. = National Fine
* See "How To Run A Lathe."

6″ IMPROVED WATER MOTOR (Continued)

OPERATIONS:

PART No. 6—SPINDLE

1. Select stock.
2. Lay off and center ends.*
3. Face ends to length.*
4. Rough turn all diameters, allowing 1/64″ stock for finishing cut.
5. Finish turn and file .436″ diameter sections for running fit in the 7/16″ reamed bearing holes in Parts No. 1 and No. 2.
6. Finish turn .439″ diameter section for press fit in reamed hole in Part No. 3.
7. Finish turn and file .374″ diameter section for slip fit into reamed hole in Part No. 5. Chamfer both ends of spindle.
8. Mark location through tapped hole in Part No. 5 and drill spot for set screw on .374″ diameter section.
9. Polish all diameters except the .439″ diameter.*

PART No. 7—COUPLING

1. Select stock.
2. Place stock in lathe chuck.*
3. Face end and chamfer outside corners.
4. Center end of stock for drilling and drill hole through stock.*
5. Bore inside diameter to be threaded and square shoulder.
6. Cut thread* fitting to threaded water faucet or standard hose coupling. Chamfer end of thread.
7. Turn work end for end in chuck and face unfinished end to length. Chamfer outside corners.

PART No. 8—COUPLING NIPPLE

1. Select stock.
2. Place stock in lathe chuck.*
3. Face and center end of stock for drilling.*
4. Drill and ream hole through stock.*
5. Press a mandrel in work and face unfinished end to length.
6. Rough turn diameters allowing 1/64″ stock for finishing cut.
7. Face shoulder to length and finish turn diameter of flange to size.
8. Finish turn and file .6885″ diameter section for press fit into reamed hole in water intake of Part No. 1—Wheel Housing.

PART No. 9—PIN

1. Press Part No. 6 into Part No. 3 as per assembly drawing.
2. Assemble water motor as per assembly drawing.
3. Drill and pin coupling nipple. (See assembly drawing.)
4. Pack stuffing box with rope asbestos treated with graphite and oil.

   NOTE A: See sketch below detail drawing of Part No. 3—Water Wheel. The arrow shows the proper direction of rotation of wheel for turning in the lathe.

*See book "How to Run a Lathe."

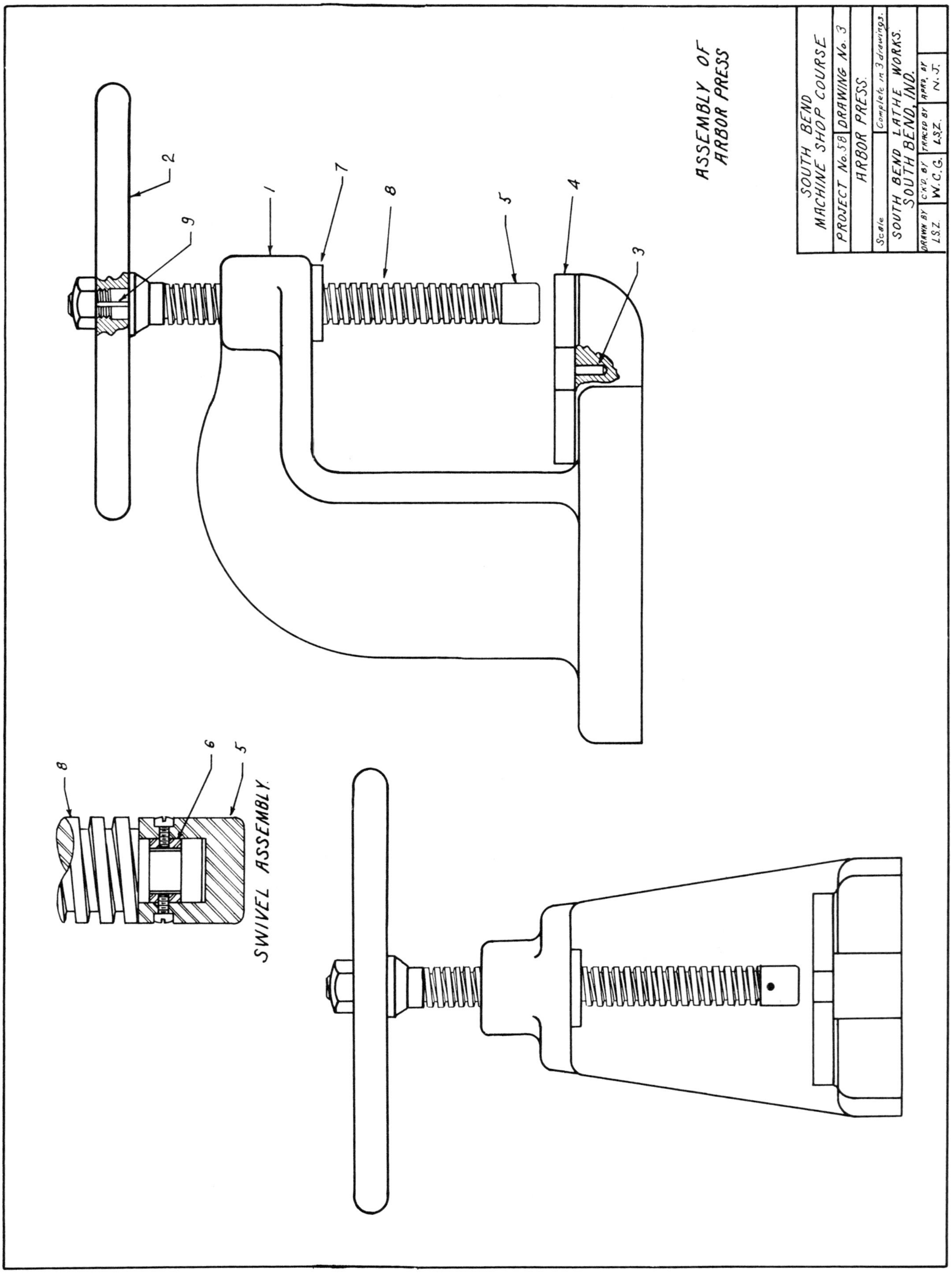
ASSEMBLY OF ARBOR PRESS
SOUTH BEND MACHINE SHOP COURSE
PROJECT No. 58
DRAWING No. 3
ARBOR PRESS.
Scale
Complete in 3 drawings.
SOUTH BEND LATHE WORKS.
SOUTH BEND, IND.
DRAWN BY L.S.Z.
CK'D. BY W.C.G.
TRACED BY L.S.Z.
APP'D. BY N.J.
SWIVEL ASSEMBLY.
1
2
3
4
5
6
7
8
9

ARBOR PRESS

MATERIAL REQUIRED:

Part No. 1—Frame. Cast iron. One required.

Part No. 2—Handwheel. Cast iron. One required.

Part No. 3—Swivel Plate Stud. Cold rolled steel 3/8" dia. x 1 7/16" long. One required.

Part No. 4—Swivel Plate. Cast iron. One required.

Part No. 5—Screw. Machinery steel 1 3/8" dia. x 3 1/2" long. One required.

Part No. 6—Screw Swivel Ring. Machinery steel 7/8" dia. x 7/16" long. One required.

Part No. 7—Nut. Cast iron. One required.

Part No. 8—Screw. Cold rolled steel 1 3/8" dia. x 14 5/16" long. One required.

Part No. 9—Key. Cold drawn steel 1/4" sq. x 1 17/32" long. One required.

Fillister Head Cap Screws—size 10-32 x 1/4" long, R.H. Two required.

OPERATIONS:

PART No. 1—FRAME

1. Remove sand and rough projections from casting.
2. Lay off, center punch and drill 25/64" hole through base of frame.

PART No. 2—HANDWHEEL

1. Remove sand and rough projections from casting.
2. Place casting in lathe chuck, with long side of hub projecting.*
3. Face and center hub for drilling.*
4. Drill and bore hole in hub allowing .005" to .010" stock for reaming to size.
5. Ream hole in hub to size.*
6. Chamfer outside corner of hub.
7. Place casting on a 1" mandrel and face unfinished side of hub to length.
8. Place casting in shaper vise and plane keyway.

PART No. 3—SWIVEL STUD

1. Select stock.
2. Place stock in drill chuck in headstock spindle.
3. Face ends to length and round corner on one end with a file.*

*See book "How to Run a Lathe."

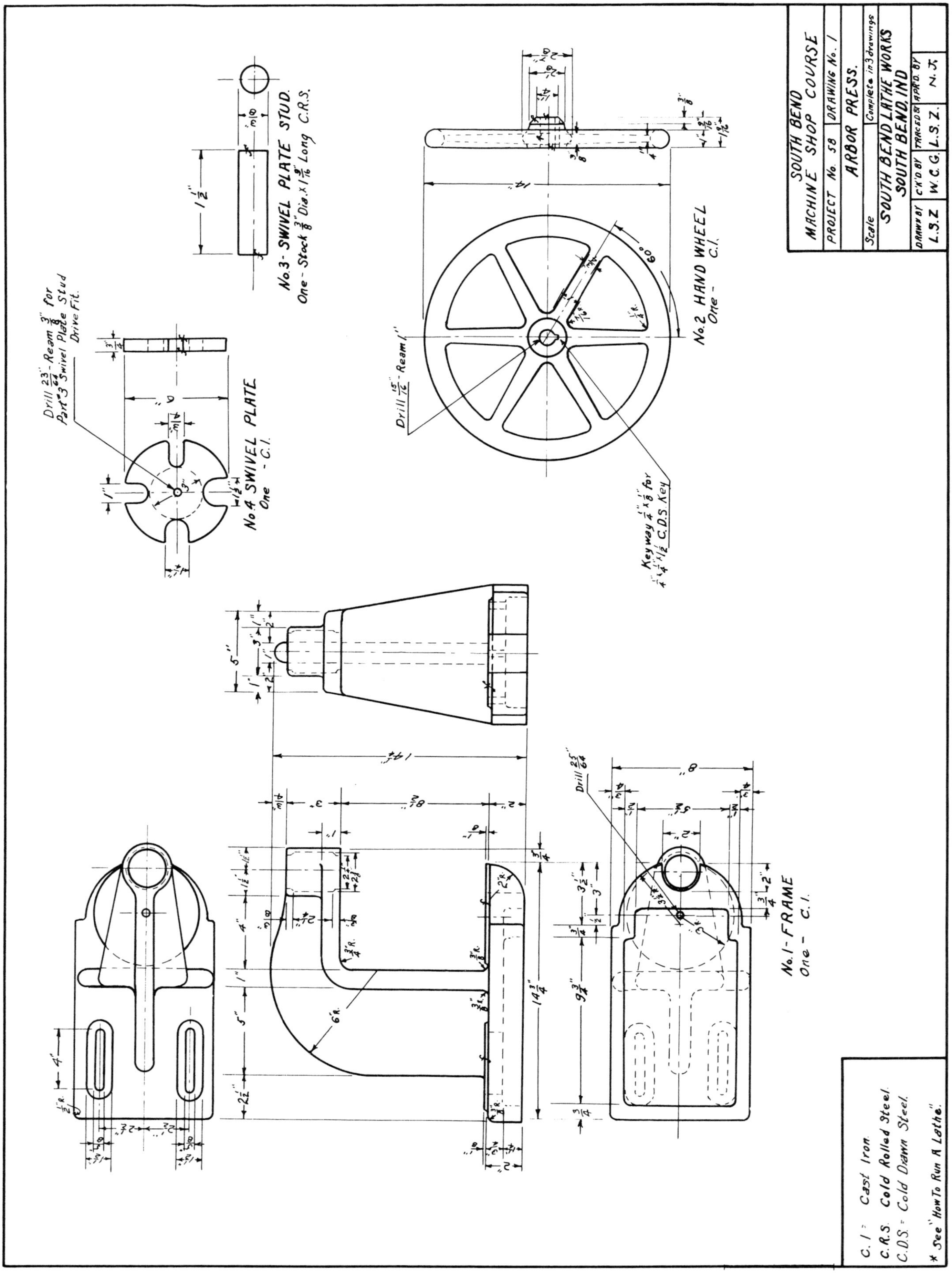
SOUTH BEND
MACHINE SHOP COURSE
PROJECT No. 58
DRAWING No. 1
ARBOR PRESS.
Scale
Complete in 3 drawings
SOUTH BEND LATHE WORKS
SOUTH BEND, IND.
DRAWN BY L.S.Z.
CK'D BY W.C.G.
TRACED BY L.S.Z.
APP'D BY N.J.
No. 1 - FRAME
One - C.I.
No. 2 HAND WHEEL
One - C.I.
No. 3 - SWIVEL PLATE STUD.
One - Stock 3/8" Dia. x 1 9/16 Long C.R.S.
No. 4 SWIVEL PLATE
One - C.I.
Drill 23/64 - Ream 3/8 for Part #3 Swivel Plate Stud Drive Fit.
Drill 15/16 - Ream 1"
Key way 1/4 x 1/8 for 1/4 x 1/4 x 1 1/2 C.D.S. Key
Drill 25/64
C.I. = Cast Iron.
C.R.S. = Cold Rolled Steel.
C.D.S. = Cold Drawn Steel.
* See "How To Run A Lathe".

ARBOR PRESS (Continued)

OPERATIONS:

PART No. 4—SWIVEL PLATE

1. Remove sand and rough projections from casting.
2. Place casting in lathe chuck.*
3. Face and center casting for drilling.*
4. Drill and ream ⅜″ hole for drive fit for Part No. 3—Swivel Stud.
5. Turn casting around in chuck and face unfinished side to width.

PART No. 5—SCREW SWIVEL

1. Select stock.
2. Place stock in lathe chuck.*
3. Face and center end of stock for drilling.*
4. Drill into end of stock to depth. Square bottom of hole with a flat drill.
5. Rough and finish turn diameter. File and polish diameter.*
6. Bore hole allowing .005″ to .010″ for reaming to size.
7. Ream hole to size. Square corner at bottom of hole with a boring tool.
8. Cut off work to length, plus 1/32″ for finishing the end.
9. Place work in lathe chuck and face unfinished end to length. Round corner with a file.
10. Lay off and center punch to drill holes for fillister head screws.
11. Drill holes and counterbore, using ¼″ counterbore with 3/16″ pilot.

PART No. 6—SWIVEL RING

1. Select stock.
2. Place stock in lathe chuck.*
3. Face and center stock for drilling.*
4. Drill and ream hole to size.*
5. Press a mandrel in work and rough and finish turn diameter, allowing .003″ for filing to size.
6. Use right and left-hand tools and face sides of ring to width without changing position of arbor between centers.
7. File diameter for slip fit in ¾″ reamed hole in Part No. 5—Screw Swivel.
8. Saw ring in two with a hacksaw.
9. The screw holes are not to be drilled until press is assembled.

*See book "How to Run a Lathe."

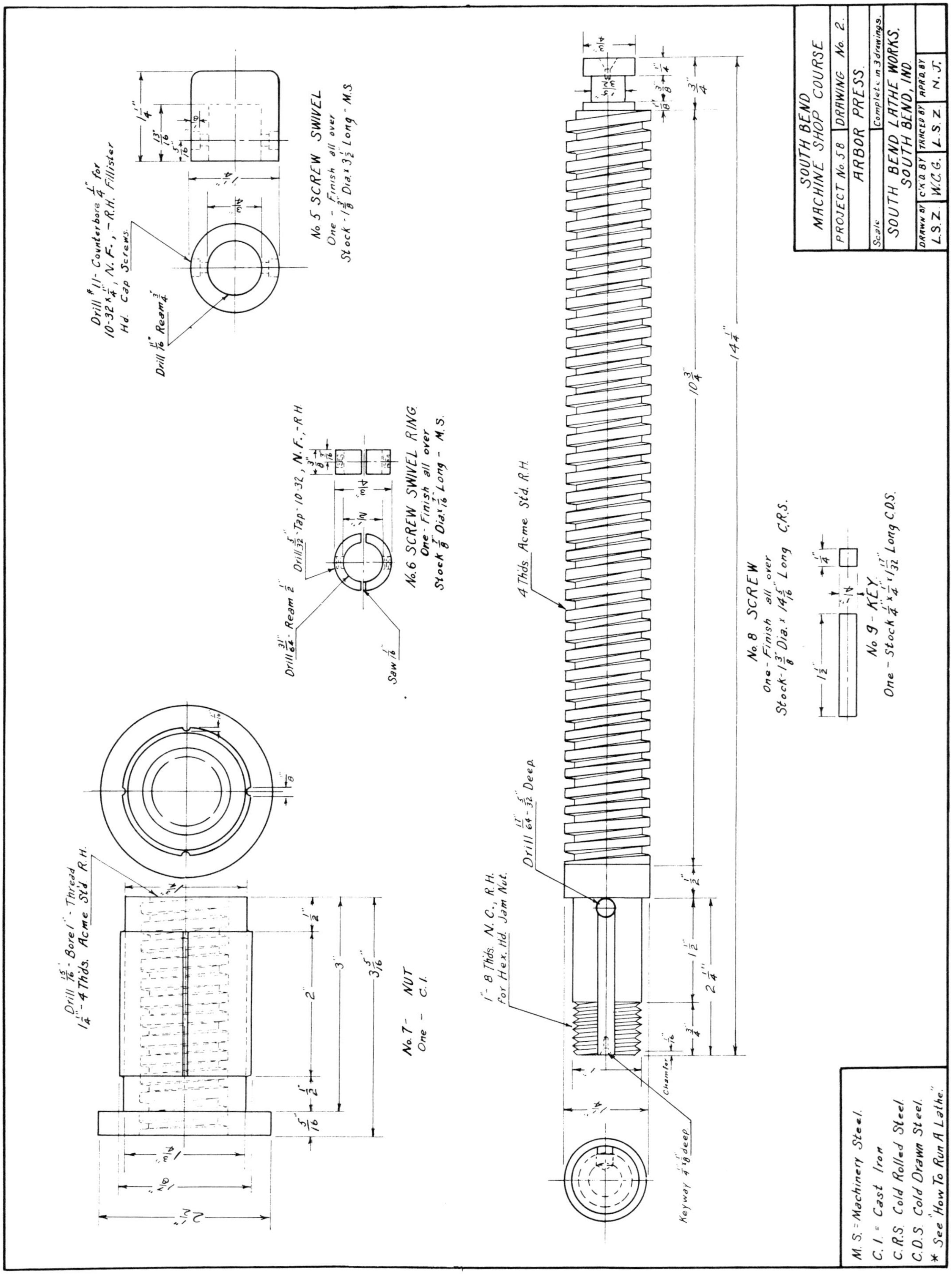
SOUTH BEND MACHINE SHOP COURSE
PROJECT No. 58
DRAWING No. 2.
ARBOR PRESS.
Scale
Complete in 3 drawings.
SOUTH BEND LATHE WORKS. SOUTH BEND, IND.
DRAWN BY L.S.Z.
C'K'D BY W.C.G.
TRACED BY L.S.Z.
APP'D BY N.J.
No. 5 SCREW SWIVEL
One - Finish all over
Stock - 1 3/8" Dia. x 3 1/2" Long - M.S.
Drill #11 - Counterbore 1/4" for 10-32 x 1/4", N.F., - R.H. Fillister Hd. Cap Screws.
Drill 11/16" Ream 3/4"
No. 6 SCREW SWIVEL RING
One - Finish all over
Stock 7/8" Dia. x 1/16" Long - M.S.
Drill 5/32" - Tap - 10-32, N.F., - R.H.
Drill 31/64 - Ream 1/2"
Saw 1/16"
No. 7 - NUT
One - C.I.
Drill 15/16" - Bore 1" - Thread 1 1/4" - 4 Thds. Acme Std. R.H.
No. 8 SCREW
One - Finish all over
Stock - 1 3/8" Dia. x 14 5/16" Long C.R.S.
4 Thds. Acme Std. R.H.
Drill 11/64 - 5/32 Deep.
1" - 8 Thds. N.C., R.H. For Hex. Hd. Jam Nut.
Keyway 1/4" x 1/8" deep
Chamfer
No. 9 - KEY
One - Stock 1/4" x 1/4" x 1 11/32" Long C.D.S.
M.S. = Machinery Steel.
C.I. = Cast Iron
C.R.S. Cold Rolled Steel
C.D.S. Cold Drawn Steel.
* See "How To Run A Lathe."

ARBOR PRESS (Continued)

OPERATIONS:

PART No. 7—NUT

1. Remove sand and rough projections from casting.
2. Place casting in lathe chuck with flange end projecting.
3. Face end and chamfer cored hole for starting drill.*
4. Drill through cored hole with a 3 or 4 lip drill.*
5. Bore hole to size.
6. Cut thread, allowing .005″ stock for tap to remove.*
7. Finish thread to size with a tap. Support and guide tap with tailstock center.*

PART No. 8—SCREW

1. Select stock.
2. Lay off and center ends.*
3. Rough turn all diameters allowing 1/64″ stock for finishing cut.
4. Face shoulder, finish turn and file 1″ diameter section for slip fit in reamed hole in Part No. 2—Handle.
5. Finish turn thread section of 1″ diameter .010″ undersize.
6. Finish turn 1¼″ diameter section to size.
7. Face shoulder, finish turn and file ¾″ diameter section for running fit in reamed hole in Part No. 5—Screw Swivel.
8. Cut 1″ diameter thread,* fitting to a 1″ National Coarse hexagon jam nut.
9. Cut Acme thread on 1¼″ diameter section,* fitting to tapped hole in Part No. 7—Nut.
10. Turn recess in ¾″ diameter section, fitting width to Part No. 6—Swivel Ring, for a running fit.
11. Lay off, center punch and drill hole for planing keyway.
12. Clamp work in shaper vise and plane keyway.

PART No. 9—KEY

1. Select stock and file key to length. Remove burrs and sharp corners with file.

ASSEMBLING INSTRUCTIONS

1. Place Part Nos. 5, 6, and 8 together as per assembly drawing.
2. Mark location of tapped holes in Part No. 6 through screw holes in Part No. 5. Remove Part No. 6 and drill and tap screw holes in location marked.
3. Assemble Part Nos. 5, 6, and 8 as per assembly drawing.
4. Assemble Part Nos. 1, 7, and 8 as per assembly drawing and babbitt Part No. 7 and Part No. 1 together.
5. Assemble Part Nos. 2, 8, and 9 as per assembly drawing and tighten the nut.

*See book "How to Run a Lathe."

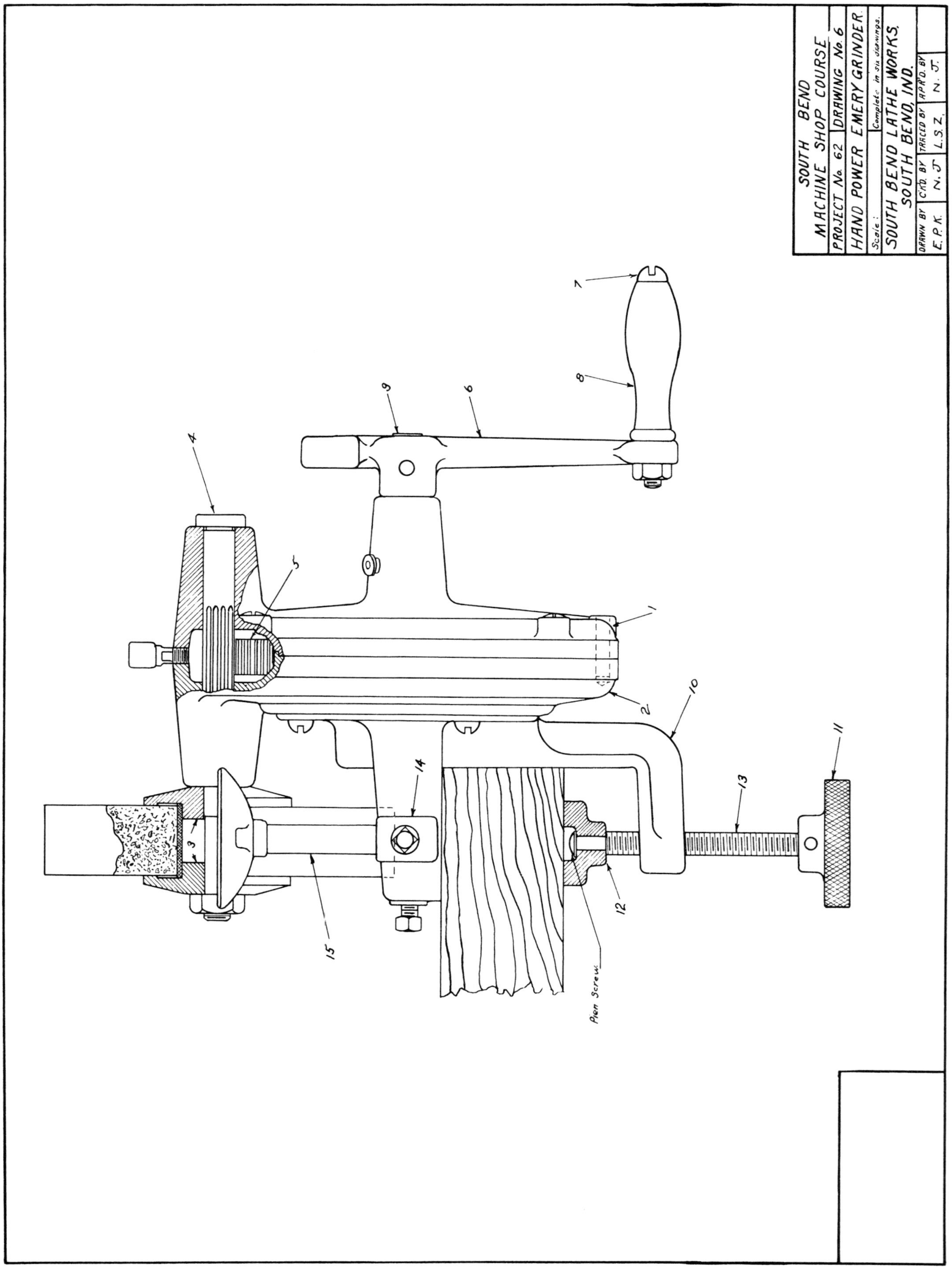
SOUTH BEND
MACHINE SHOP COURSE
PROJECT No. 62
DRAWING No. 6
HAND POWER EMERY GRINDER.
Scale:
Complete in six drawings.
SOUTH BEND LATHE WORKS,
SOUTH BEND, IND.
DRAWN BY E.P.K.
C'K'D. BY N.J.
TRACED BY L.S.Z.
APP'D. BY N.J.
Plain Screw
1
2
3
4
5
6
7
8
9
10
11
12
13
14
15

HAND POWER EMERY GRINDER

MATERIAL REQUIRED:

Part No. 1—Gear Case—Right Half. Cast iron. One required.

Part No. 2—Gear Case—Left Half. Cast iron. One required.

Part No. 3—Flange. Cast iron. Two required.

Part No. 4—Spindle. Cold rolled steel 13/16" dia. x 5 27/32" long. One required.

Part No. 5—Bull Gear. Cast iron. One required.

Part No. 6—Crank. Cast iron. One required.

Part No. 7—Handle Bolt. Cold rolled steel 1/2" dia. x 5" long. One required.

Part No. 8—Handle. Hardwood 1" sq. x 4" long. One required.

Part No. 9—Bull Gear Shaft. Cold rolled steel 1/2" dia. x 5 3/8" long. One required.

Part No. 10—Clamp Bracket. Cast iron. One required.

Part No. 11—Knob. Machinery steel 1 7/8" dia. x 3" long. One required.

Part No. 12—Swivel. Cast iron. One required.

Part No. 13—Cold rolled steel 3/8" dia. x 3 13/16" long. One required.

Part No. 14—"T" Rest Bracket. Cast iron. One required.

Part No. 15—"T" Rest. Cast iron. One required.

Tucker Oiler. No. 22 Style G. One required.

Ball Oiler Size 1/4" No. O-A. One required.

Round Head Machine Screws 1/4" dia. x 1/2" long. National Coarse R.H. Three required.

Fillister Head Cap Screws 3/16" dia. x 3/4" long. National Coarse R.H. Four required.

Square Head Set Screw 1/4" dia. x 1/4" National Coarse R.H. One required.

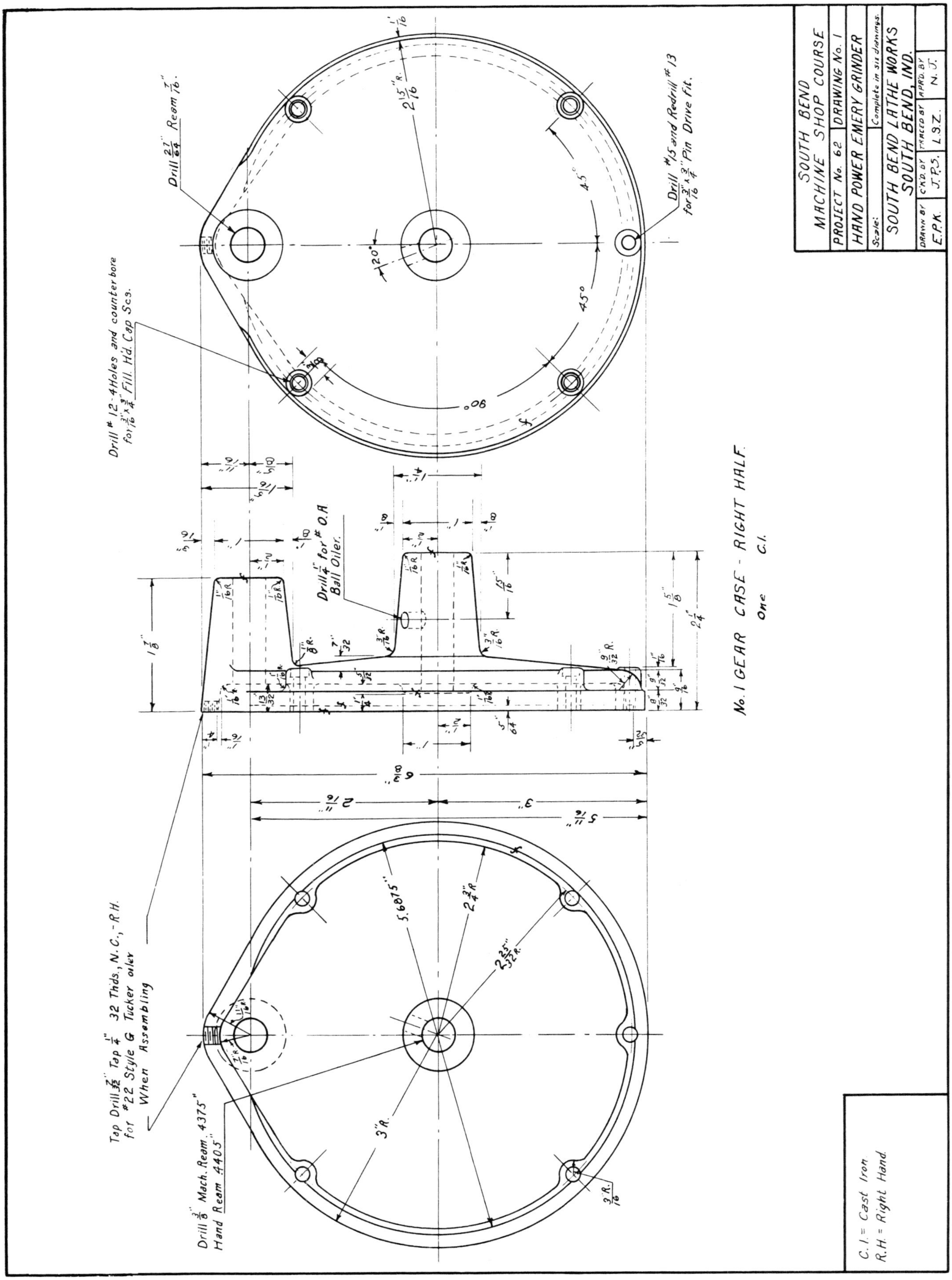
SOUTH BEND MACHINE SHOP COURSE
PROJECT No. 62
DRAWING No. 1
HAND POWER EMERY GRINDER
Scale:
Complete in six drawings.
SOUTH BEND LATHE WORKS SOUTH BEND, IND.
DRAWN BY E.P.K.
CK'D. BY J.P.S.
TRACED BY L.S.Z.
APP'D. BY N.J.
No. 1 GEAR CASE - RIGHT HALF. C.I.
One
Drill 21/64 Ream 7/16
Drill #15 and Redrill #13 for 3/16 x 3/4 Pin Drive fit.
Drill #12 - 4 Holes and counter bore for 3/16 x 3/4 Fill. Hd. Cap Scs.
Drill 1/4 for #0 A Ball Oiler.
Tap Drill 7/32 Tap 1/4 32 Thds., N.C., - R.H. for #22 Style G Tucker oiler When Assembling
Drill 3/8 Mach. Ream .4375 Hand Ream .4405
C.I. = Cast Iron.
R.H. = Right Hand.

HAND POWER EMERY GRINDER (Continued)

OPERATIONS:

PART No. 1—GEAR CASE—RIGHT HALF

1. Remove sand and rough projections from casting.
2. Place casting in lathe chuck* with irregular side out.
3. Face end of long boss to length.
4. Center casting for drilling and drill hole through long boss.*
5. Bore hole, allowing .005″ to .010″ stock for reaming to size.
6. Machine ream hole to standard diameter.*
7. Press a mandrel in casting. Face surface and bore recess for centering the two halves of the gear case when joined together.
8. Face short boss inside of casting to depth.
9. Lay off and center punch location to all holes to be drilled.
10. Drill hole in long boss for ball oiler. Drill and counterbore 3/16″ screw holes. Use 1/4″ counterbore with 3/16″ pilot.

    NOTE: Do not drill the rest of the holes until instructed.

PART No. 2—GEAR CASE—LEFT HALF

1. Remove sand and rough projections from casting.
2. Place casting in lathe chuck* with irregular side out.
3. Face end of short boss to length.* Face rim of clamp bracket recess to height. (See footnote A).
4. Center casting for drilling and drill hole through short boss.*
5. Bore hole allowing .005″ to .010″ for reaming to size.
6. Machine ream hole to standard diameter.*
7. Press a mandrel in casting. Face surface and turn tongue for centering the two halves of the gear case when joined together. Face tongue to height.
8. Lay off and center punch location of holes to be tapped in clamp bracket recess.
9. Place the two halves together as per assembly drawing.
10. Drill dowel pin hole in both parts. Remove Part No. 1 and redrill pin hole in Part No. 2 for slip fit.
11. Drive the dowel pin in Part No. 1 and place the gear case back together.
12. Spot location of 3/16″ screw holes onto Part No. 2 and drill screw holes with tap drill to depth.
13. Drill holes in clamp bracket recess with tap drill.
14. Tap 3/16″ screw holes, using holes in Part No. 1 to guide tap. Insert and tighten the screws.
15. Center punch, drill and tap gear case for Tucker Oiler No. 22 Style G.

    NOTE: Do not tap 1/4″ holes in clamp bracket recess until instructed.
16. Drill and ream hole through both halves of gear cast for Part No. 4—Spindle Pinion.
17. Spot face ends of spindle pinion bearing boss to length. Use 1 1/8″ counterbore with 7/16″ pilot.

    NOTE: The 7/16″ hole through center of case is to be hand reamed for running fit for part No. 9—Bull Gear Shaft, when grinder is assembled.

*See book "How to Run a Lathe."

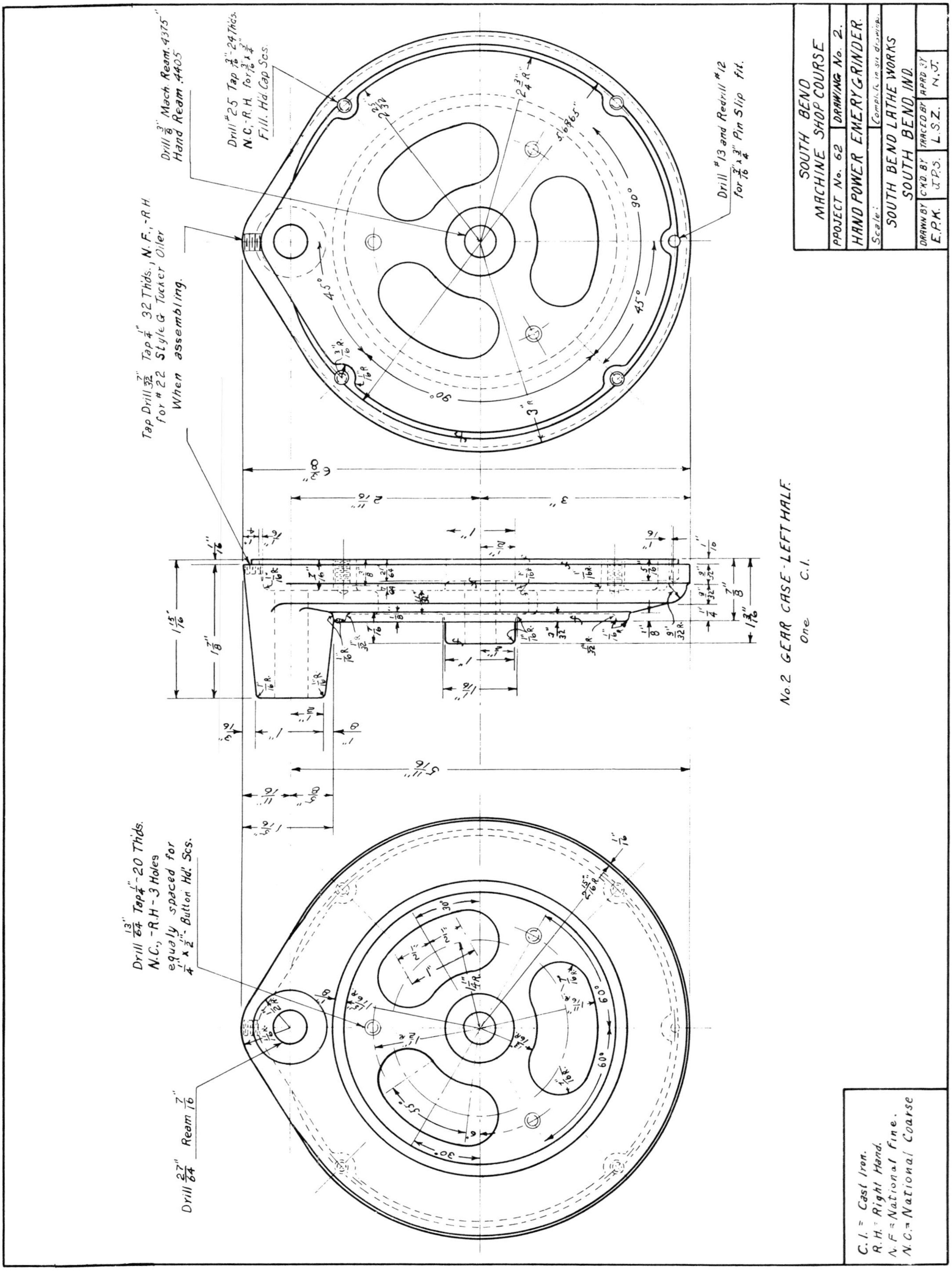
SOUTH BEND MACHINE SHOP COURSE
PROJECT No. 62
DRAWING No. 2.
HAND POWER EMERY GRINDER.
Scale:
SOUTH BEND LATHE WORKS
SOUTH BEND IND.
DRAWN BY E.P.K.
C'K'D. BY J.P.S.
TRACED BY L.S.Z.
APP'D BY N.J.
No. 2 GEAR CASE - LEFT HALF.
One C.I.
C.I. = Cast Iron.
R.H. = Right Hand.
N.F. = National Fine.
N.C. = National Coarse
Drill 3/8" Mach. Ream .4375" Hand Ream .4405
Drill #25 Tap 3/16"-24 Thds. N.C.-R.H. for 3/16" x 1/2" Fill. Hd. Cap Scs.
Drill #13 and Redrill #12 for 3/16" x 3/4" Pin Slip fit.
Tap Drill 7/32" Tap 1/4" 32 Thds., N.F.,-R.H. for #22 Style G Tucker Oiler When assembling.
Drill 13/64" Tap 1/4"-20 Thds. N.C., -R.H.-3 Holes equally spaced for 1/4" x 1/2" Button Hd. Scs.
Drill 27/64" Ream 7/16"

HAND POWER EMERY GRINDER (Continued)

OPERATIONS:

PART No. 3—FLANGE

1. Remove sand and rough projections from casting.
2. Place casting in lathe chuck* with beveled side out.
3. Face* and center casting for drilling.*
4. Drill and ream hole in casting.
5. Press a mandrel in casting and face both unfinished shoulders to length.
6. Rough turn diameters and angle allowing 1/64″ stock for finishing cut.
7. Finish turn diameters and angle to size. File and polish large diameter and beveled side of flange. Undercut face of large shoulder.

PART No. 4—SPINDLE

1. Select stock.
2. Lay off and center ends.*
3. Face ends to length.*
4. Rough turn all diameters, allowing 1/64″ stock for finishing cut.
5. Finish turn head section and round outside corner. File inside corner and file and polish diameter.
6. Finish turn, file and polish .4365″ diameter section for running fit in reamed hole in gear case.
7. Finish turn short .3765″ diameter section for press fit into 3/8″ reamed hole in Part No. 3—Flange.
8. Finish turn .374″ diameter section for slip fit into reamed hole in Part No. 3—Flange.
9. Finish turn section to be threaded .010″ undersize.
10. Cut thread* fitting to 3/8″ National Coarse Jam Nut. Chamfer end of thread.
11. Mill pinion gear teeth in .4365″ diameter section.

PART No. 5—BULL GEAR

1. Remove sand and rough projections from casting.
2. Place casting in lathe chuck.*
3. Rough face one side and center casting for drilling.*
4. Drill hole through hub of casting.
5. Bore hole allowing .005″ to .010″ stock for reaming to size.
6. Ream hole to standard size.
7. Press a mandrel in casting and rough face unfinished side of casting.
8. Finish face both sides of casting to width.
9. Rough and finish turn diameter.
10. Mill gear teeth.

PART No. 6—CRANK

1. Remove sand and rough projections from casting.
2. Lay off and center punch location of holes to be drilled.
3. Drill and machine ream hole through long boss.
4. Drill and tap hole through small boss at end of crank.

   NOTE: Do not drill taper hole until instructed. The reamed hole through the long boss is to be hand reamed for slip fit on Part No. 9—Bull Gear Shaft when grinder is assembled.

*See book "How to Run a Lathe."

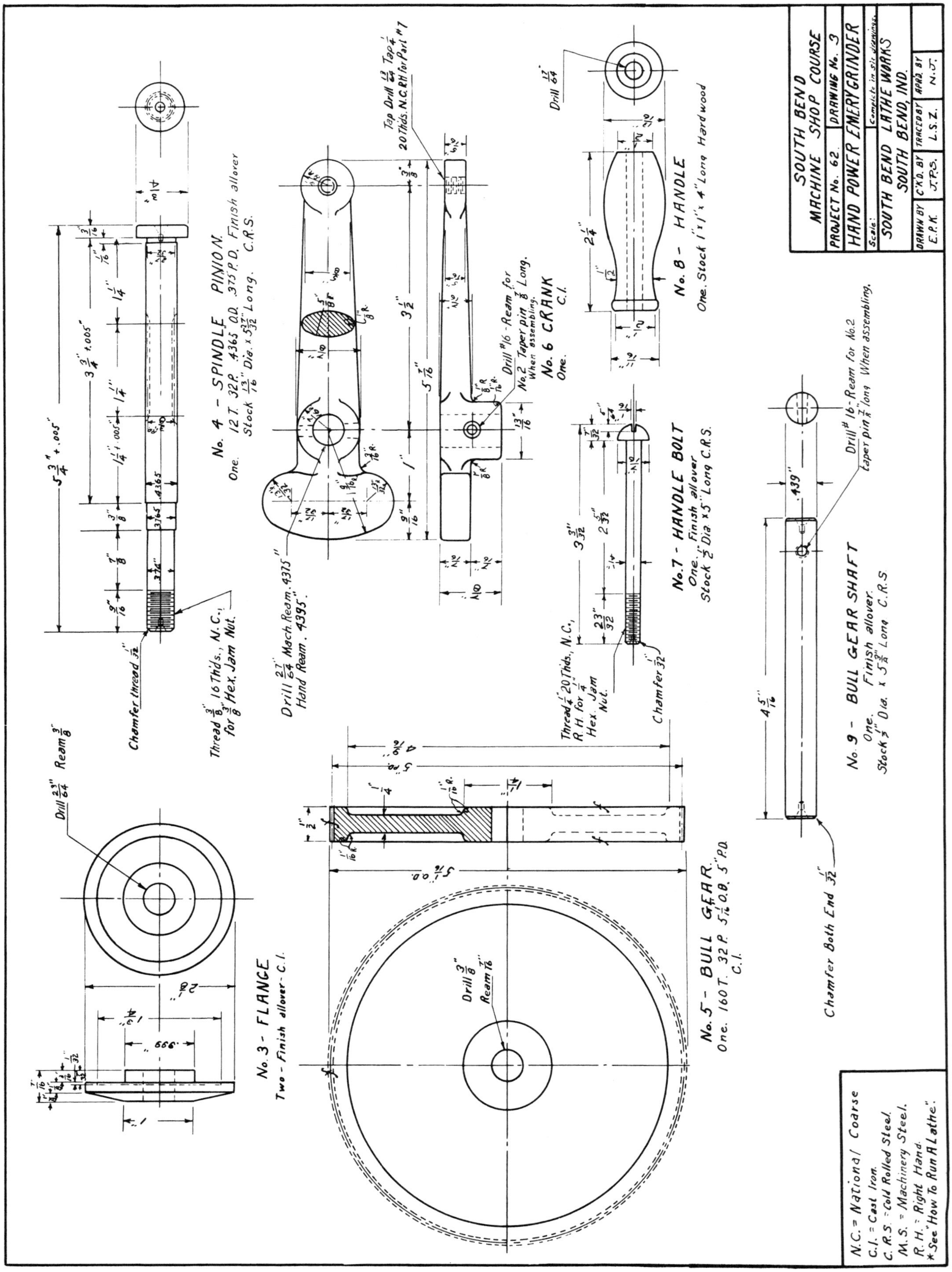
SOUTH BEND MACHINE SHOP COURSE
PROJECT No. 62.
DRAWING No. 3
HAND POWER EMERY GRINDER
Scale:
Complete in six drawings.
SOUTH BEND LATHE WORKS SOUTH BEND, IND.
DRAWN BY E.P.K.
C'K'D. BY J.P.S.
TRACED BY L.S.Z.
APP'D BY N.J.
No. 4 - SPINDLE PINION.
One. 12 T. 32 P. .4365 O.D. .375 P.D. Finish allover
Stock 13/16" Dia. x 5 27/32 Long. C.R.S.
Thread 3/8" 16 Th'ds., N.C.-1, for 3/8" Hex. Jam Nut.
Chamfer thread 1/32"
No. 6 CRANK C.I.
One.
Drill 27/64" Mach. Ream .4375" Hand Ream .4395".
Drill #16 - Ream for No. 2 Taper pin 7/8 Long. When assembling.
Tap Drill 13/64 Tap 1/4 20 Th'ds. N.C. RH for Part #7
No. 8 - HANDLE
One. Stock 1"x1"x4" Long Hardwood
Drill 17/64"
No. 7 - HANDLE BOLT
One. Finish allover
Stock 1/2" Dia x 5" Long C.R.S.
Thread 1/4" 20 Th'ds. N.C.-1 R.H. for 1/4" Hex. Jam Nut.
Chamfer 1/32"
No. 9 - BULL GEAR SHAFT
One. Finish allover.
Stock 1/2" Dia. x 5 3/8" Long C.R.S
Drill #16 - Ream for No. 2 taper pin 7/8" long When assembling.
Chamfer Both End 1/32"
No. 5 - BULL GEAR.
One. 160 T. 32 P. 5 1/16 O.D. 5" P.D. C.I.
Drill 3/8" Ream 7/16"
No. 3 - FLANGE
Two - Finish allover - C.I.
Drill 23/64 Ream 3/8
N.C. = National Coarse
C.I. = Cast Iron.
C.R.S. = Cold Rolled Steel.
M.S. = Machinery Steel.
R.H. = Right Hand.
* See "How To Run A Lathe".

HAND POWER EMERY GRINDER (Continued)

OPERATIONS:

PART No. 7—HANDLE BOLT

1. Select stock.
2. Place stock in lathe chuck.*
3. Face end.* Center drill and support end of work with tailstock center.*
4. Rough turn diameters, allowing 1/64″ stock for finishing cut.
5. Finish turn diameters to size.
6. Turn section to be threaded .010″ undersize.
7. Cut thread* fitting to tapped hole in Part No. 6—Crank. Chamfer end of thread.
8. Cut off work to length, plus 1/32″ stock for finishing the end.
9. Place work in lathe chuck* with head end of work projecting.
10. Face end of head section to length. Round end of head. File and polish.
11. Saw screw driver slot in head section.

PART No. 8—HANDLE

1. Select stock.
2. Place stock in lathe chuck.
3. Face end.* Support end of work with cup center in tailstock spindle.
4. Rough turn handle.
5. Finish turn handle to shape. Round end and sand surface smooth.
6. Drill into work with 17/64″ drill 2⅜″ deep.
7. Cut off work to length.

PART No. 9—BULL GEAR SHAFT

1. Select stock.
2. Lay off and center ends.*
3. Face ends to length*, plus 1″ stock for attaching the lathe dog.
4. Rough turn diameter allowing 1/64″ stock for finishing cut.
5. Finish turn, file and polish diameter for press fit into reamed hole in Part No. 5—Bull Gear. Chamfer end.
6. Saw off the surplus stock allowed for attaching lathe dog allowing 1/16″ for finishing end of work.
7. Place work in lathe chuck* and face unfinished end to length.

   NOTE: Do not drill taper hole until instructed.

PART No. 10—CLAMP BRACKET

1. Remove sand and rough projections from casting.
2. Chalk face of flange and transfer location of ¼″ screw holes, marking through drill holes in clamp bracket recess of Part No. 2—Gear Case, Left Half.
3. Lay off and center punch location of all other holes to be drilled.
4. Drill all holes and tap for screws where specified. Ream ½″ hole for "T" Rest Bracket.

*See book "How to Run a Lathe."

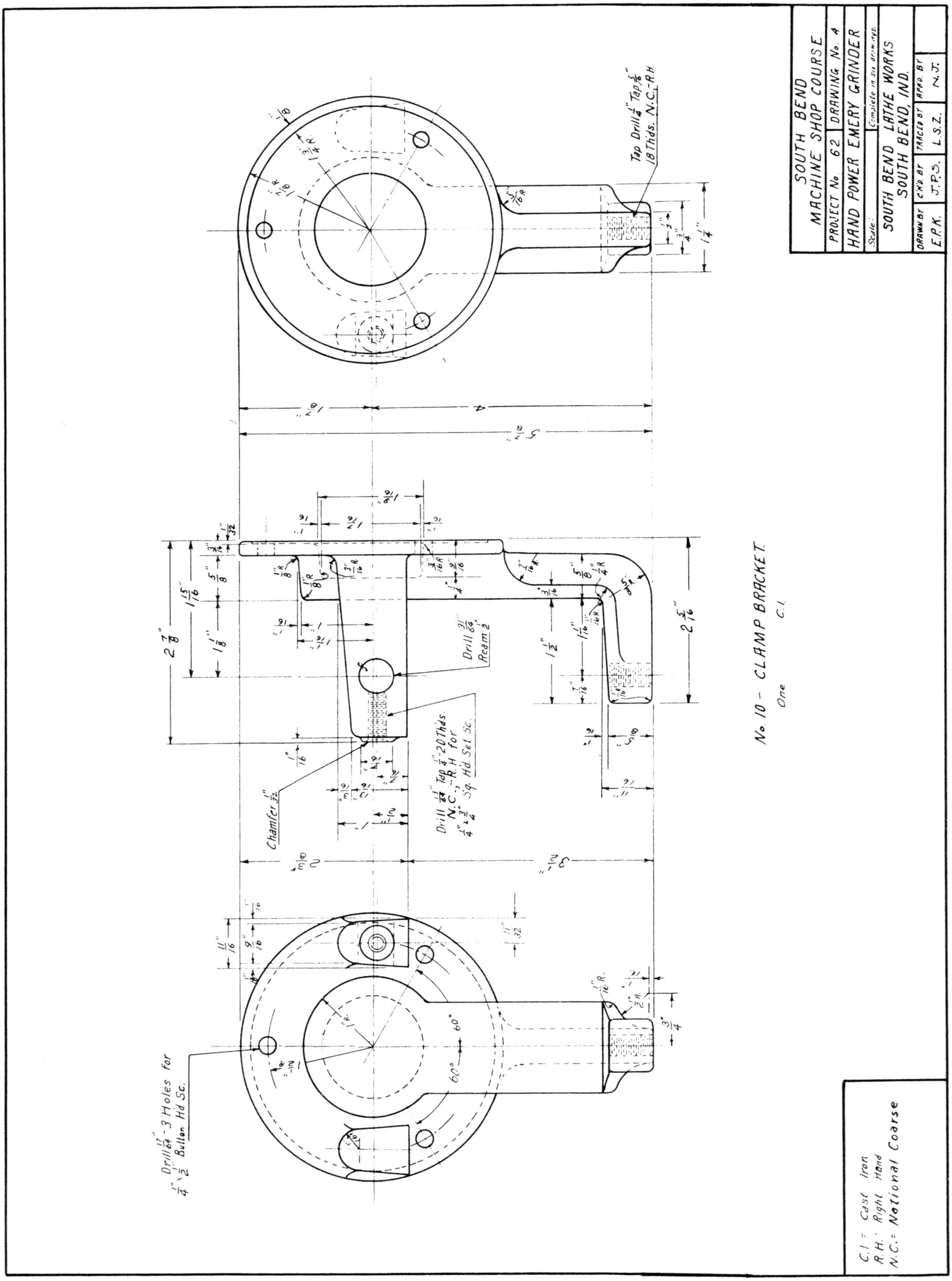
SOUTH BEND
MACHINE SHOP COURSE
PROJECT No. 62 DRAWING No. 4
HAND POWER EMERY GRINDER
Scale: Complete in six drawings
SOUTH BEND LATHE WORKS
SOUTH BEND, IND.
DRAWN BY E.P.K. CK'D BY J.P.S. TRACED BY L.S.Z. APP'D. BY N.J.
No. 10 - CLAMP BRACKET. C.I.
One
Tap Drill 1/4" Tap 5/16" 18 Thds. N.C.-R.H.
Drill 31/64" Ream 1/2"
Chamfer 1/32"
Drill 13/64" Tap 1/4"-20 Thds. N.C.-R.H. for 1/4" x 3/4" Sq. Hd. Set Sc.
Drill 17/64" - 3 Holes for 1/4" x 1/2" Button Hd Sc.
C.I. = Cast Iron
R.H. = Right Hand
N.C. = National Coarse

HAND POWER EMERY GRINDER (Continued)

OPERATIONS:

PART No. 11—KNOB

1. Select stock.
2. Place stock in lathe chuck.*
3. Face end.* Center drill and support end of work with tailstock center.
4. Rough turn diameters, allowing 1/64″ stock for finishing cut.
5. Finish turn large diameter section.*
6. Knurl large diameter section.*
7. Finish turn small diameter section, face shoulder of large diameter section and form fillet in corner.
8. Round end of small diameter section with file. File and polish machined surfaces.
9. Drill hole in hub to depth and ream to size.
10. Cut off work to length, plus 1/32″ stock for finishing the end.
11. Place work in lathe chuck and face unfinished end to length. File corner and polish face.*
    NOTE: Do not drill taper pin hole until instructed.

PART No. 12—SWIVEL

1. Remove sand and rough projections from casting.
2. Place casting in lathe chuck with hub projecting.
3. Face hub to length and center casting for drilling.*
4. Drill hole through casting.

PART No. 13—SCREW

1. Select stock.
2. Lay off and center ends.*
3. Rough turn large diameter section, allowing 1/64″ stock for finishing cut.
4. Finish turn large diameter section to size.
5. Finish turn section to be threaded .010″ undersize.
6. Cut thread,* fitting to 5/16″ tapped hole in Part No. 10—Clamp Bracket.
7. Turn pilot section to size and face shoulder to length.
   NOTE: Do not drill taper pin hole until instructed.

PART No. 14—"T" REST BRACKET

1. Remove sand and rough projections from casting.
2. Lay off and center ends.*
3. Face end to length.*
4. Rough turn diameter allowing 1/64″ stock for finishing cut.
5. Finish turn diameter allowing .003″ for filing to size.
6. File and polish diameter to size. Chamfer end.
7. Lay off, center punch and drill and ream hole through "T" boss.
8. Drill and tap set screw hole between centers in lathe.

PART No. 15—"T" REST

1. Remove sand and rough projections from casting.
2. Lay off and center ends.*
3. Face end to length.*
4. Rough turn diameter allowing 1/64″ stock for finishing cut.
5. Finish turn, file and polish diameter to size. Chamfer end.
   NOTE A: The flange of the clamp recess is faced to provide a finished surface for leveling the work on the drill press.

ASSEMBLING INSTRUCTIONS

1. Hand ream 7/16″ hole in Part No. 6—Crank for light press fit for Part No. 9—Bull Gear Shaft and press shaft in handle.
2. Drill through both parts, ream hole for taper pin. Remove handle and re-ream hole for slip fit on Part No. 9—Bull Gear Shaft.
3. Press Part No. 13—Screw, into reamed hole in Part No. 11—Knob. Drill through both parts and ream for taper pin.
4. Assemble grinder as per assembly drawing.

*See book "How to Run a Lathe."

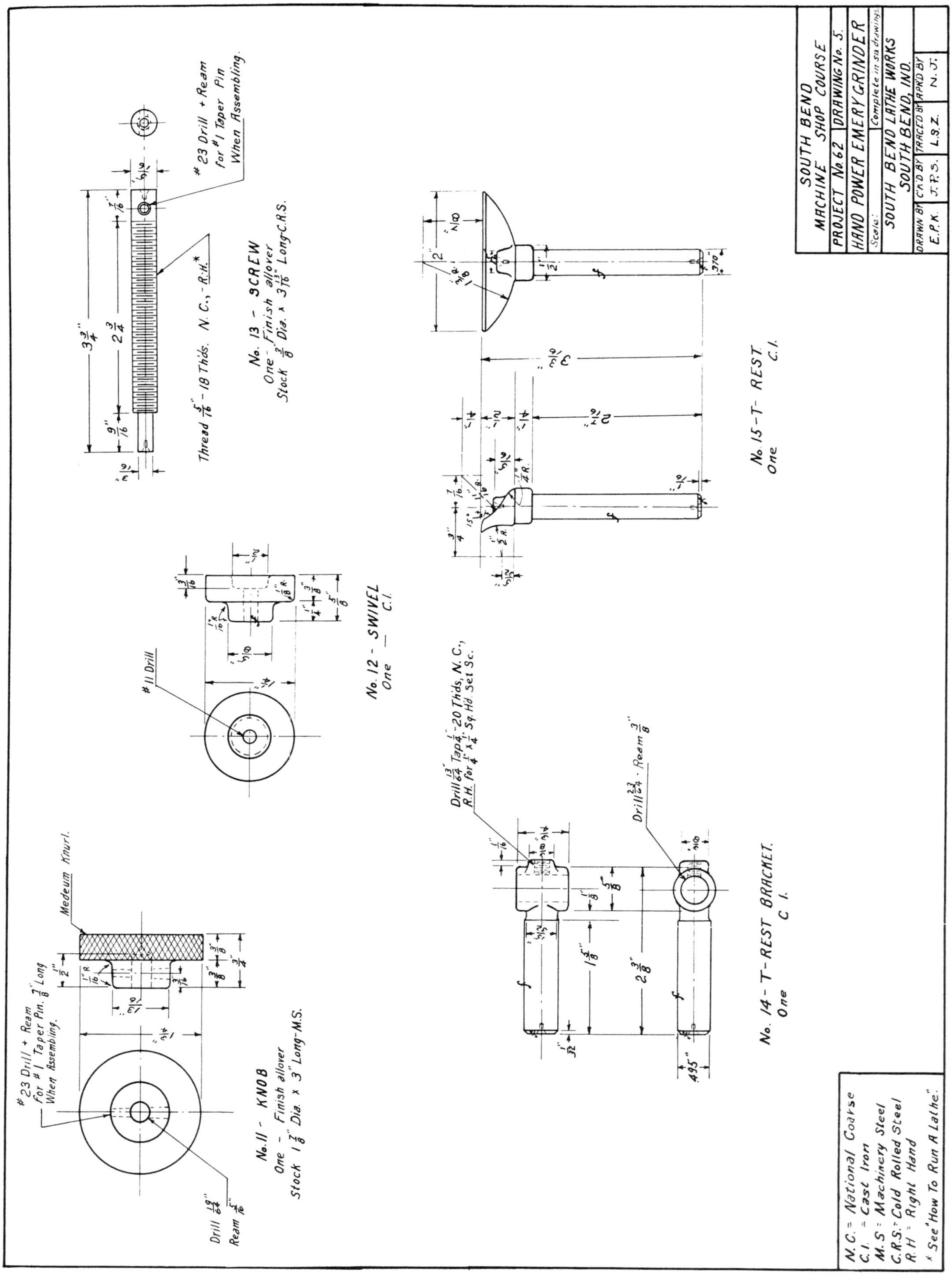
SOUTH BEND MACHINE SHOP COURSE
PROJECT No. 62
DRAWING No. 5.
HAND POWER EMERY GRINDER
Scale:
Complete in six drawings.
SOUTH BEND LATHE WORKS SOUTH BEND, IND.
DRAWN BY E.P.K.
CH'K'D BY J.P.S.
TRACED BY L.S.Z.
APP'RD BY N.J.
No. 13 - SCREW
One - Finish allover
Stock 3/8" Dia. x 3 13/16" Long-C.R.S.
Thread 5/16" - 18 Thds. N.C., - R.H.*
#23 Drill + Ream for #1 Taper Pin When Assembling.
No. 15 - T - REST
One C.I.
No. 12 - SWIVEL
One — C.I.
#11 Drill
Drill 13/64 Tap 1/4" - 20 Thds, N.C., R.H. for 1/4" x 1/4" Sq. Hd. Set Sc.
Drill 23/64, Ream 3/8"
No. 14 - T-REST BRACKET.
One C.I.
Medeum Knurl.
#23 Drill + Ream For #1 Taper Pin. 7/8" Long When Assembling.
No. 11 - KNOB
One - Finish allover
Stock 1 7/8" Dia. x 3" Long-M.S.
Drill 19/64"
Ream 5/16"
N.C. = National Coarse
C.I. = Cast Iron
M.S. = Machinery Steel
C.R.S. = Cold Rolled Steel
R.H. = Right Hand
* See "How To Run A Lathe".

SCREWS, NUTS, ETC.	AM'T.
3/16"-24 X 3/8" Fill. Hd. Cap Scs.	2
1/4"-20 X 5/8" Fill. Hd. Cap Scs.	8
1/4"-20 X 3/4" Fill. Hd. Cap Scs.	8
1/16" X 1/4" Straight C.R.S. Pins.	2
5/16"-18 X 1/2" Hdless. Set Sc.	1
#8-32 X 1/4" Button Hd. Sc.	1
#8-32 Hex. Nut	1
#6-32 X 5/8 Button Hd. Scs.	2
#10-32 X 1/2" Button Hd. Sc.	1
#3 X 1 3/8 Taper Pin.	1
#4 X 1 7/8" Taper Pins.	2
3/16" X 9/16" Straight C.R.S. Pin.	1
#6-32 X 1/4" Button Hd. Scs.	4
#10-32 X 1 1/8" Button Hd. Scs.	2
#10-32 Square Nuts.	2
ACCESSORIES.	**AM'T.**
1/2" Champion X Spark Plug.	1
Greese Cups with 1/8" Pipe Thds.	2
1/4" Pipe Ell.	1
1/4" Pipe Plug	1
1/4" Street Elbow.	2
1/4" Pipe Union	1
3/16"-24 X 5/8" Fill. Hd. Cap Sc.	1
1/8 X 1 1/2" Pipe Nipples.	2
1/8 Pipe Couplings.	2
1/2" X 3" Pipe Nipple	1

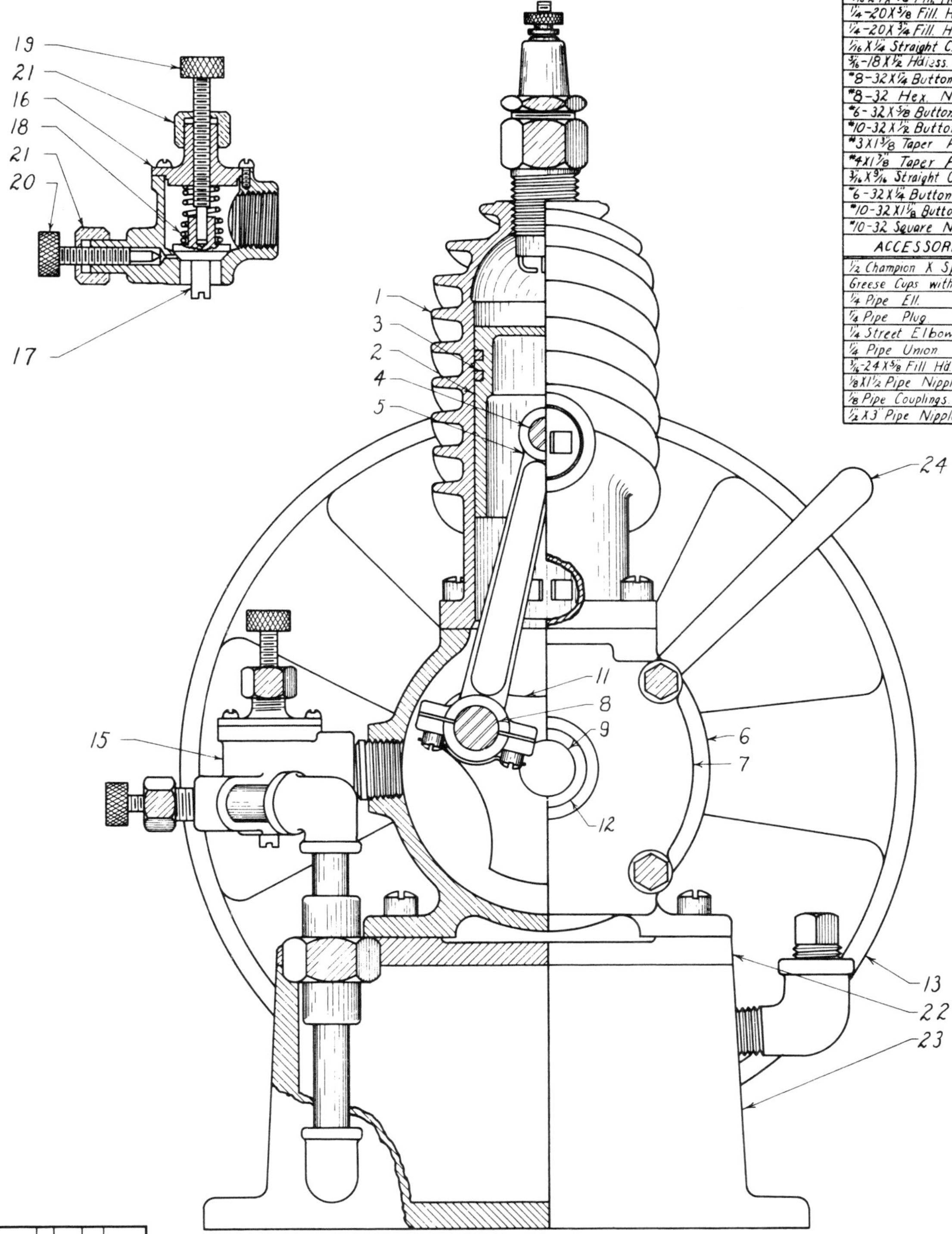

ASSEMBLY OF 1/4 H.P. GAS. ENGINE - VERTICAL

1 CYLINDER
2 PISTON
3 PISTON RING
4 WRIST PIN
5 CONNECTING ROD
6 CRANK CASE
7 CR. CASE COVER
8 CRANK PIN
9 CRANK SHAFT
10 CRANK SHAFT
11 CR. SHAFT DISC
12 CR. SHAFT BUSHING
13 FLYWHEEL
14 PULLEY
15 MIXING VALVE
16 MIX. VALVE CAP
17 AIR VALVE
18 AIR VALVE SPRING
19 AIR ADJ. SCREW
20 FUEL ADJ. SCREW
21 PACKING NUT
22 FUEL TANK COVER
23 FUEL TANK
24 SPARK LEVER
25 SPARK BRUSH
26 SP. BR. INSULATOR
27 SP. BR. SCREW INS.
28 CONTACT POINT
29 STARTING CRANK
30 COMMUTATOR
31 MUFFLER (INSIDE)
32 MUFFLER (OUTSIDE)

SOUTH BEND
MACHINE SHOP COURSE
PROJECT No. 70 | DRAWING No. 8
1/4 H.P. GAS. ENGINE - VERTICAL
Scale: | Complete in 9 Drawings
SOUTH BEND LATHE WORKS
SOUTH BEND, IND.

DRAWN	C'K'D.	TRACED	APR'D.
N.D.J.	J.P.S.	J.H.	N.J.

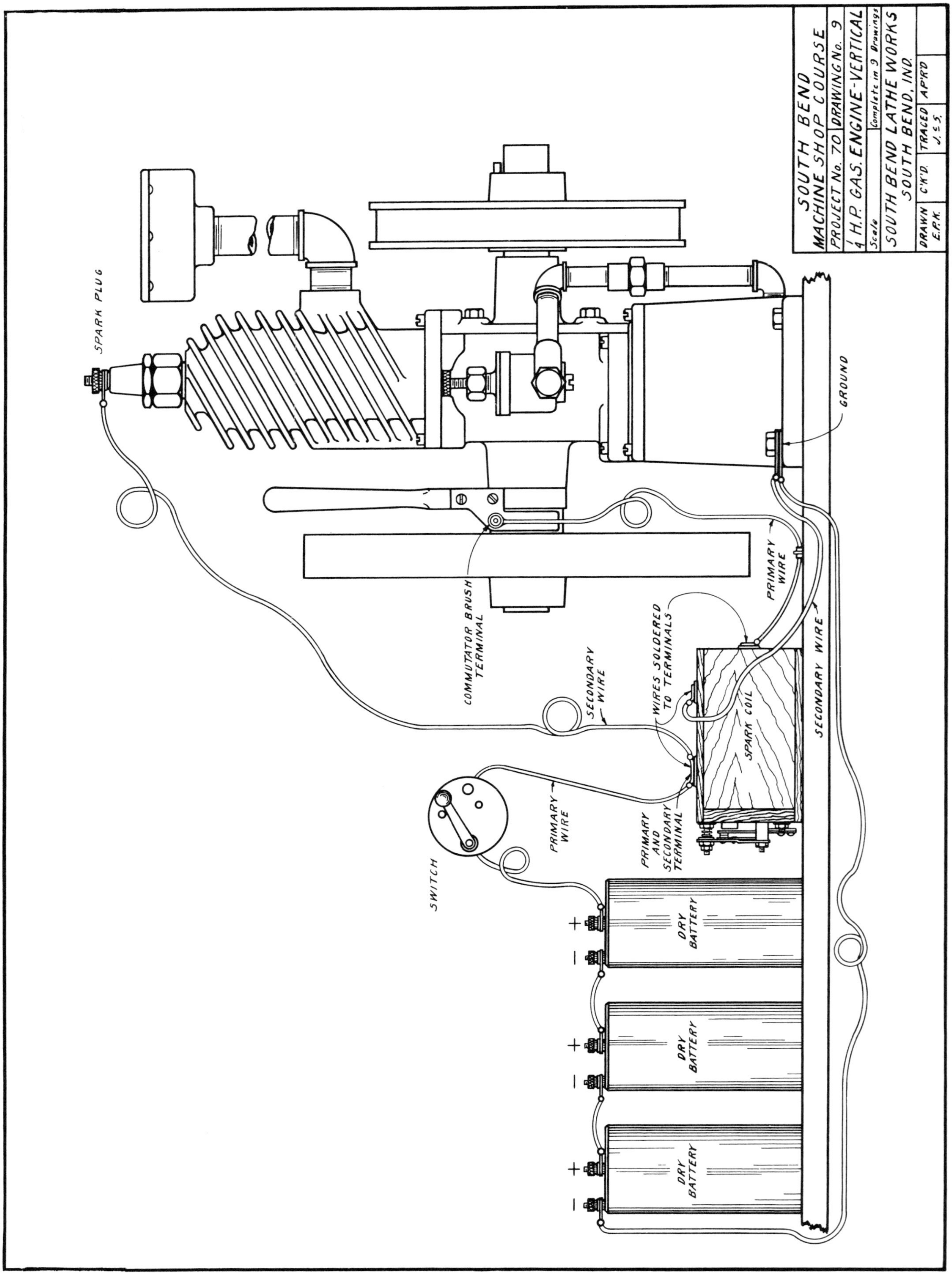
SOUTH BEND
MACHINE SHOP COURSE
PROJECT No. 70
DRAWING No. 9
1/4 H.P. GAS ENGINE-VERTICAL
Scale
Complete in 9 Drawings
SOUTH BEND LATHE WORKS
SOUTH BEND, IND.
DRAWN E.P.K.
C'K'D.
TRACED J.S.S.
AP'R'D
SPARK PLUG
GROUND
COMMUTATOR BRUSH TERMINAL
PRIMARY WIRE
WIRES SOLDERED TO TERMINALS
SECONDARY WIRE
SPARK COIL
SECONDARY WIRE
PRIMARY WIRE
PRIMARY AND SECONDARY TERMINAL
SWITCH
+
-
DRY BATTERY
+
-
DRY BATTERY
+
-
DRY BATTERY

¼ H.P. VERTICAL AIR-COOLED GASOLINE ENGINE

MATERIAL REQUIRED:

Part No. 1—Cylinder. Cast iron. One required.

Part No. 2—Piston. Cast iron. One required.

Part No. 3—Piston Ring. Cast iron. Two required.

Part No. 4—Wrist Pin. Machinery steel 7/16″ dia. x 3″ long. One required.

Part No. 5—Connecting Rod and Cap. Cast bronze. One of each required.

Part No. 6—Crankcase. Cast iron. One required.

Part No. 7—Crankcase Cover. Cast iron. One required.

Part No. 8—Crank Pin. Machinery steel ⅝″ dia. x 3″ long. One required.

Part Nos. 9 and 10—Crankshaft. Machinery steel ¾″ dia. x 8 15/16″ long. One required.

Part No. 11—Crankshaft Disc. Cold rolled steel 2 13/16″ dia. x ½″ long. Two required.

Part No. 12—Crankshaft Bushing. Cast bronze. Two required.

Part No. 13—Flywheel. Cast iron. One required.

Part No. 14—Pulley. Cast iron. One required.

Part No. 15—Mixing Valve. Cast brass. One required.

Part No. 16—Mixing Valve Cap. Cast brass. One required.

Part No. 17—Air Valve. Cast brass. One required.

Part No. 18—Air Valve Spring. Piano wire No. 18 gauge x 12″ long, required for each spring.

Part No. 19—Air Adjusting Screw. Bar brass 9/16″ dia. x 4″ long. One required.

Part No. 20—Gasoline Adjusting Screw. Bar brass 9/16″ dia. x 3″ long. One required.

Part No. 21—Packing Nut. Bar brass ½″ hexagon x 2″ long. One required to make two nuts.

Part No. 22—Fuel Tank Cover. Cast iron. One required.

Part No. 23—Fuel Tank. Cast iron. One required.

Part No. 24—Spark Lever. Cast iron. One required.

Part No. 25—Spark Brush. Sheet brass 1/32″ x ⅞″ x 2⅛″ long. One required.

Part No. 26—Spark Brush Insulator. Fibre 3/16″ x 7/16″ x 1 1/16″ long. One required.

Part No. 27—Spark Brush Screw Insulator. Fibre ½″ dia. x 1½″ long. One required for two insulators.

Part No. 28—Contact Point. Bar brass 7/16″ dia. x 2″ long. One required.

Part No. 29—Starting Crank. Cast iron. One required.

Part No. 30—Commutator. Fibre 1½″ dia. x 9/16″ long. One required.

Part No. 31—Muffler (inside). Cast iron. One required.

Part No. 32—Muffler (outside). Cast iron. One required.

COLD ROLLED STEEL PINS

1/16″ dia. x ¼″ long. Two required.

3/16″ dia. x 9/16″ long. One required.

HEADLESS SET SCREWS

Cup point 5/16″—18 x ½″ long. One required.

FILLISTER HEAD CAP SCREWS

No. 12—24 x ⅜″ long. Two required.

No. 12—24 x ⅝″ long. One required.

¼″—20 x ⅝″ long. Eight required.

¼″—20 x ¾″ long. Eight required.

BUTTON HEAD SCREWS

No. 6—32 x ¼″ long. Four required.

No. 6—32 x ⅝″ long. Two required.

No. 8—32 x ¼″ long. One required.

No. 10—32 x ½″ long. One required.

No. 10—32 x 1⅛″ long. Two required.

HEXAGON NUTS

No. 8—32. One required.

No. 10—32. Two required.

TAPER PINS

No. 3 x 1⅜″ long. One required.

No. 4 x 1⅞″ long. Two required.

ACCESSORIES

½″ Champion X Spark Plug. One required.

Grease cups—Empress 000. Two required.

¼″ Pipe elbow. One required.

¼″ Pipe plug. One required.

¼″ Street elbow. Two required.

¼″ Pipe union. One required.

⅛″ Pipe nipple 1½″ long. Two required.

⅛″ Pipe coupling. Two required.

½″ Pipe nipple. One required.

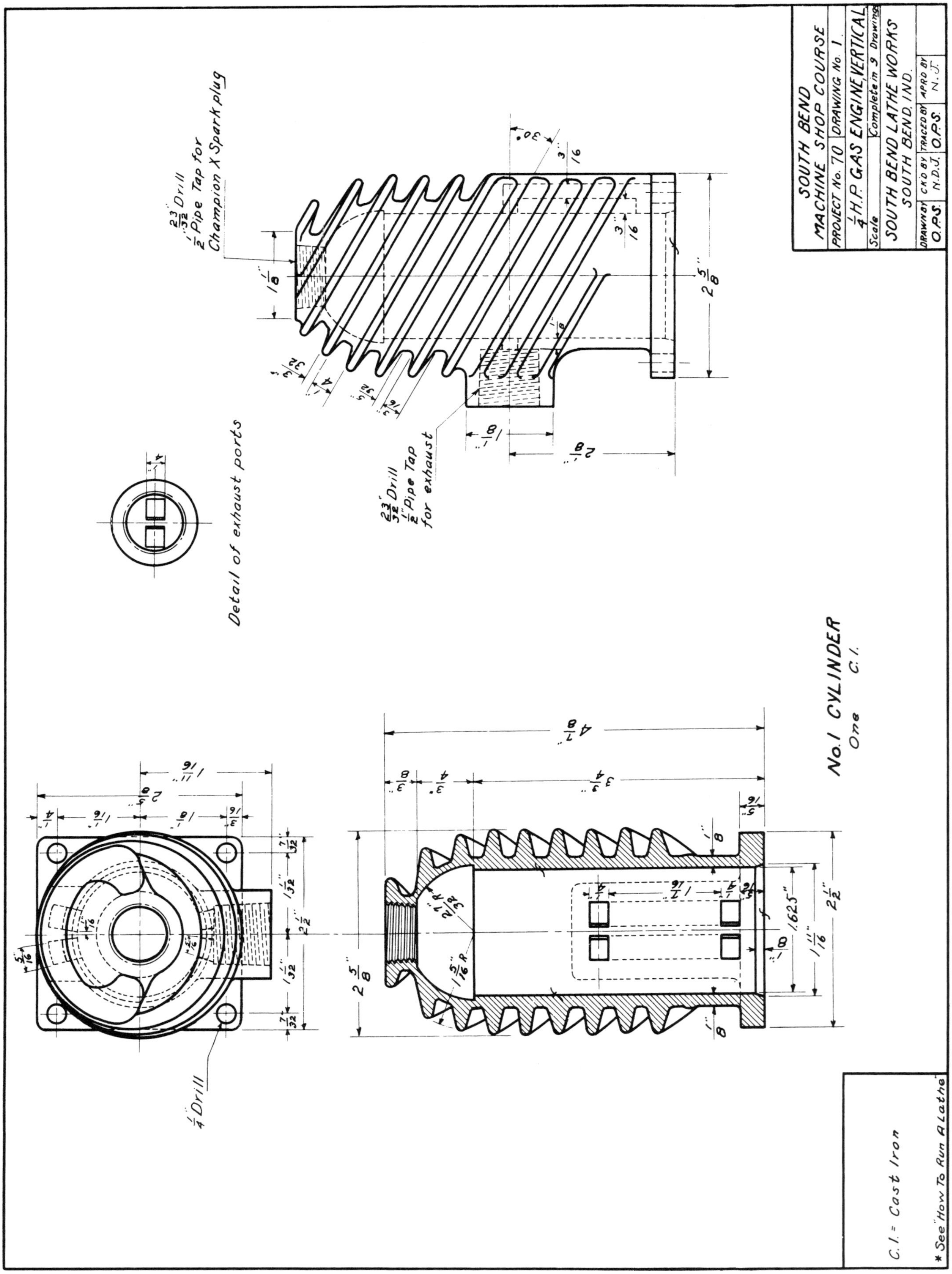
SOUTH BEND
MACHINE SHOP COURSE
PROJECT No. 70
DRAWING No. 1
1/4 H.P. GAS ENGINE VERTICAL
Scale
Complete in 3 Drawings
SOUTH BEND LATHE WORKS
SOUTH BEND, IND.
DRAWN BY O.P.S.
C.K.D BY N.D.J.
TRACED BY O.P.S.
APR'D BY N.J.
23/32" Drill
1/2" Pipe Tap for
Champion X Spark plug
23/32" Drill
1/2" Pipe Tap
for exhaust
Detail of exhaust ports
No. 1 CYLINDER
One C.I.
1/4" Drill
.1625"
C.I. = Cast Iron
* See "How To Run A Lathe"

1/4 H.P. VERTICAL AIR-COOLED GASOLINE ENGINE (Continued)

OPERATIONS:

JOB No. 1—PART No. 1—CYLINDER

1. Chuck casting.
2. Face bottom flange.
3. Bore cylinder.
4. Chamfer corner.
5. Ream to size.

JOB No. 2—PART No. 2—PISTON

1. Chuck casting with skirt projecting.
2. Face end of skirt.
3. Bore inside diameter at bottom of skirt.
4. Rough drill wrist pin hole.
5. Place piston on a ring piston adapter.
6. Rough turn diameter.
7. Rough turn ring grooves.
8. Finish turn diameter and ring grooves.
9. Bore and ream wrist pin hole.
10. Drill dowel pin holes in ring grooves.
11. Spot face inside ends of wrist pin bosses.

JOB No. 3—PART No. 3—PISTON RING

1. Chuck casting.
2. Bore and ream to size.
3. Press a mandrel in casting.
4. Face ends.
5. Rough turn diameter.
6. Cut off rings, allowing finish on sides.
7. Face sides to width.
8. Saw.
9. Finish turn diameter.

NOTE: File notches for dowel pins in assembly.

JOB No. 4—PART No. 4—WRIST PIN

1. Chuck stock.
2. Face end.
3. Rough turn diameter.
4. Finish turn diameter.
5. File and polish.
6. Round end.
7. Cut off, allowing finish on end.
8. Chuck work.
9. Face unfinished end to length.
10. Round end.

JOB No. 5—PART No. 5—CONNECTING ROD

1. Separate the connecting rod cap from the rod with a 1/32" saw.
2. Cut shims to fit between connecting rod and cap.
3. Drill cap.
4. Drill and tap connecting rod for cap.
5. Insert shims and attach cap with screws.
6. Drill and ream connecting rod for crank pin and wrist pin.
7. Drill oil hole.
8. Saw binding screw boss.
9. Drill and tap for binding screw.
10. Face wrist pin boss to length.
11. Face crank pin bearing to length.

JOB No. 6—PART No. 7—CRANKCASE COVER

1. Chuck casting with long end of bearing boss projecting.
2. Face end of bearing boss.
3. Drill, bore and ream.
4. Press a mandrel in casting.
5. Face unfinished end of bearing boss to length.
6. Face shoulder of flange.
7. Turn diameter of flange projection.
8. Drill and tap.

NOTE: Holes for attaching the cylinder are to be drilled in assembly.

JOB No. 7—PART No. 22—FUEL TANK COVER

1. Plane bottom.
2. Plane top.
3. Drill and tap.*

*See book "How to Run a Lathe."

PROJECT No. 47

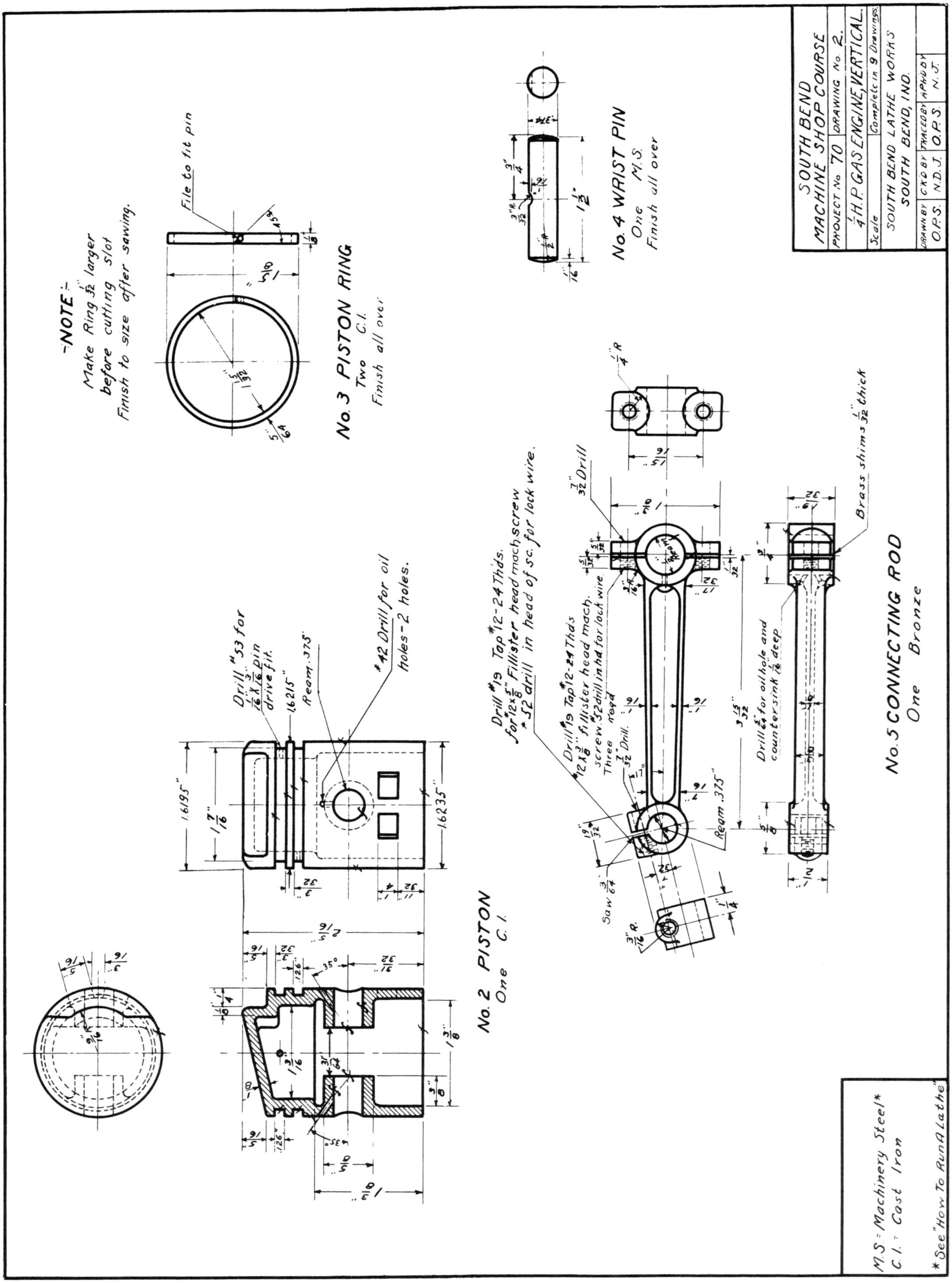

¼ H.P. VERTICAL AIR-COOLED GASOLINE ENGINE (Continued)

OPERATIONS:

JOB No. 8—PART No. 23—FUEL TANK

1. Plane top.
2. Drill and tap.

JOB No. 9—PART No. 31—MUFFLER (INSIDE)

1. Drill

JOB No. 10—PART No. 32—MUFFLER (OUTSIDE)

1. Drill and tap.

JOB No. 11—PART No. 6—CRANKCASE

1. Chuck casting with open side out.
2. Face hub and surface for cover.
3. Bore case for cover.
4. Drill, bore and ream bearing hole.
5. Press a mandrel in casting.
6. Face outside end of hub.
7. Turn diameter of hub.
8. Face shoulder to length.
9. Drill and tap.
10. Plane bottom.
11. Assemble gear case and cover.
12. Plane or mill surface for cylinder.

JOB No. 12—PART No. 12—BUSHING

1. Chuck casting.
2. Drill, bore, and ream.
3. Press a mandrel in casting.
4. Face ends to length.
5. Rough turn diameter.
6. Finish turn diameter.
7. File and polish.
8. Drill oil hole.
9. Plane or chip oil groove.

JOB No. 13—PART No. 13—FLYWHEEL

1. Chuck casting.
2. Face hub and side of rim.
3. Drill, bore, and ream.
4. Press a mandrel in casting.
5. Face unfinished end of hub and side of rim.
6. Turn diameter.
7. File and polish.

NOTE: The taper pin hole to be drilled in assembly.

JOB No. 14—PART No. 14—DRIVING PULLEY

1. Chuck casting.
2. Face hub and side of rim.
3. Drill, bore, and ream.
4. Press a mandrel in casting.
5. Face unfinished end of hub and side of rim.
6. Turn diameter.
7. Turn groove for belt.
8. File and polish.
9. Drill and tap for set screw.
10. Drill hub for 3/16″ pin.

JOB No. 15—PART No. 29—STARTING CRANK

1. Drill.
2. File finish cam.

JOB No. 16—PART No. 11—CRANKSHAFT DISK

NOTE: Perform operations No. 1 to No. 6 on both disks.

1. Chuck stock.
2. Drill, bore, and ream .625″ diameter hole.
3. Press a mandrel in work.
4. Face sides to width.
5. Turn diameter.
6. Place face of disks together.
7. Press a .625″ diameter plug into .625″ diameter holes in disks to keep holes in line.
8. Lay off and punch center of .500″ diameter hole on the top disk.
9. Clamp work on large face plate of lathe.
10. True up center point for boring .500″ diameter hole with a center test indicator.
11. Drill, bore, and ream .500″ diameter hole through both disks.
12. Lay off to remove stock from cut-away sections.
13. Drill, chip, grind and file, or mill stock from cut-away sections.

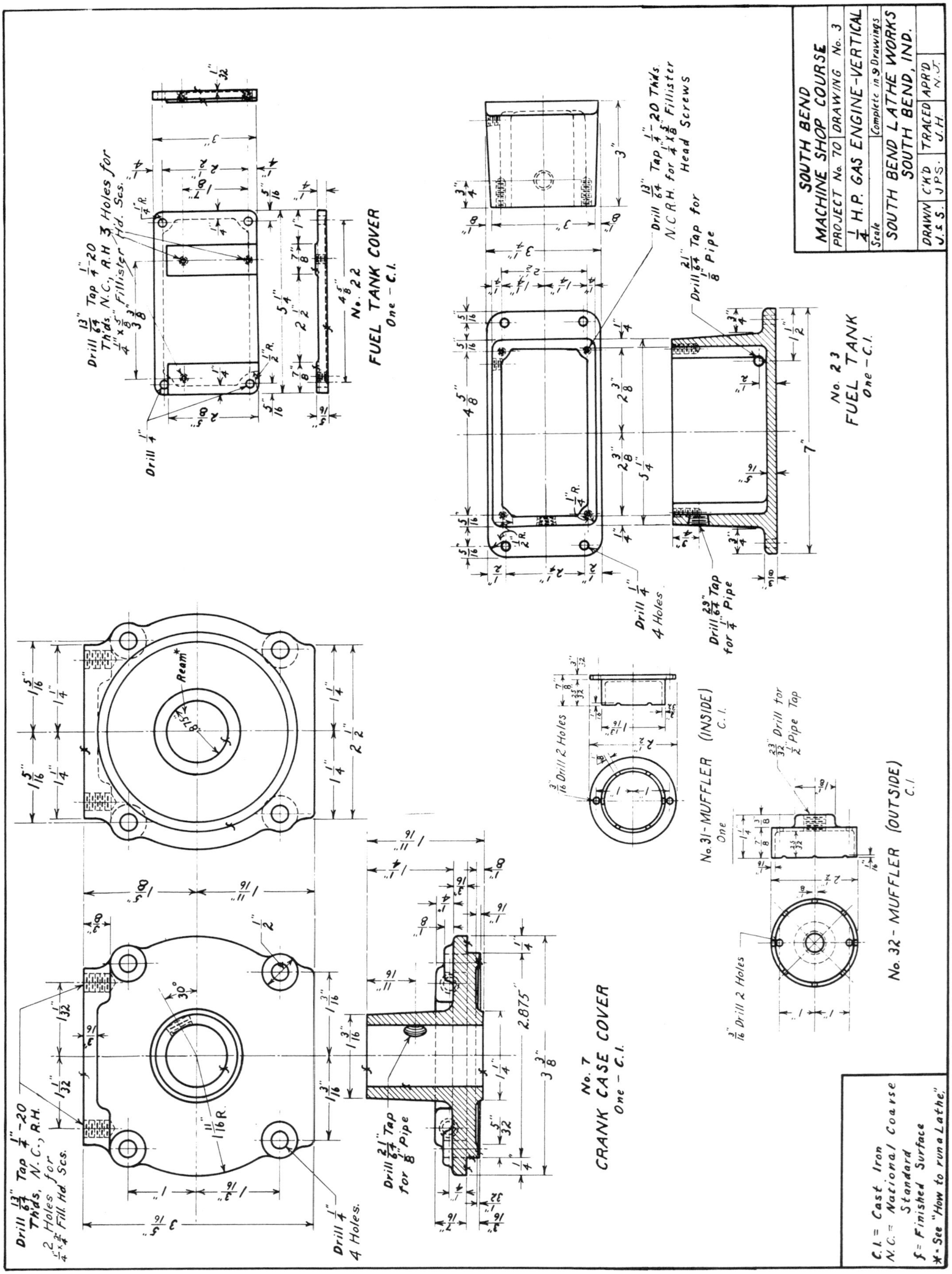
SOUTH BEND MACHINE SHOP COURSE
PROJECT No. 70 DRAWING No. 3
1/4 H.P. GAS ENGINE-VERTICAL
Scale
Complete in 9 Drawings
SOUTH BEND LATHE WORKS SOUTH BEND, IND.
DRAWN J.S.S.
C'K'D J.P.S.
TRACED J.H.
APR'D N.J.
Drill 13/64" Tap 1/4"-20 Th'ds. N.C., R.H. 3 Holes for 1/4"x5/8" Fillister Hd. Scs.
Drill 1/4"
No. 22 FUEL TANK COVER One - C.I.
Drill 13/64" Tap 1/4"-20 Th'ds. N.C.R.H. for 1/4"x5/8" Fillister Head Screws
Drill 21/64" Tap for 1/8" Pipe
Drill 1/4" 4 Holes
Drill 23/64" Tap for 1/4" Pipe
No. 23 FUEL TANK One - C.I.
Ream*
Drill 13/64" Tap 1/4"-20 Th'ds, N.C., R.H. 2 Holes for 1/4"x1/2" Fill. Hd. Scs.
Drill 1/4" 4 Holes.
Drill 21/64" Tap for 1/8" Pipe
No. 7 CRANK CASE COVER One - C.I.
3/16 Drill 2 Holes
No. 31-MUFFLER (INSIDE) One C.I.
23/32 Drill for 1/2" Pipe Tap
3/16 Drill 2 Holes
No. 32-MUFFLER (OUTSIDE) C.I.
C.I. = Cast Iron
N.C. = National Coarse Standard
f = Finished Surface
* = See "How to run a Lathe."

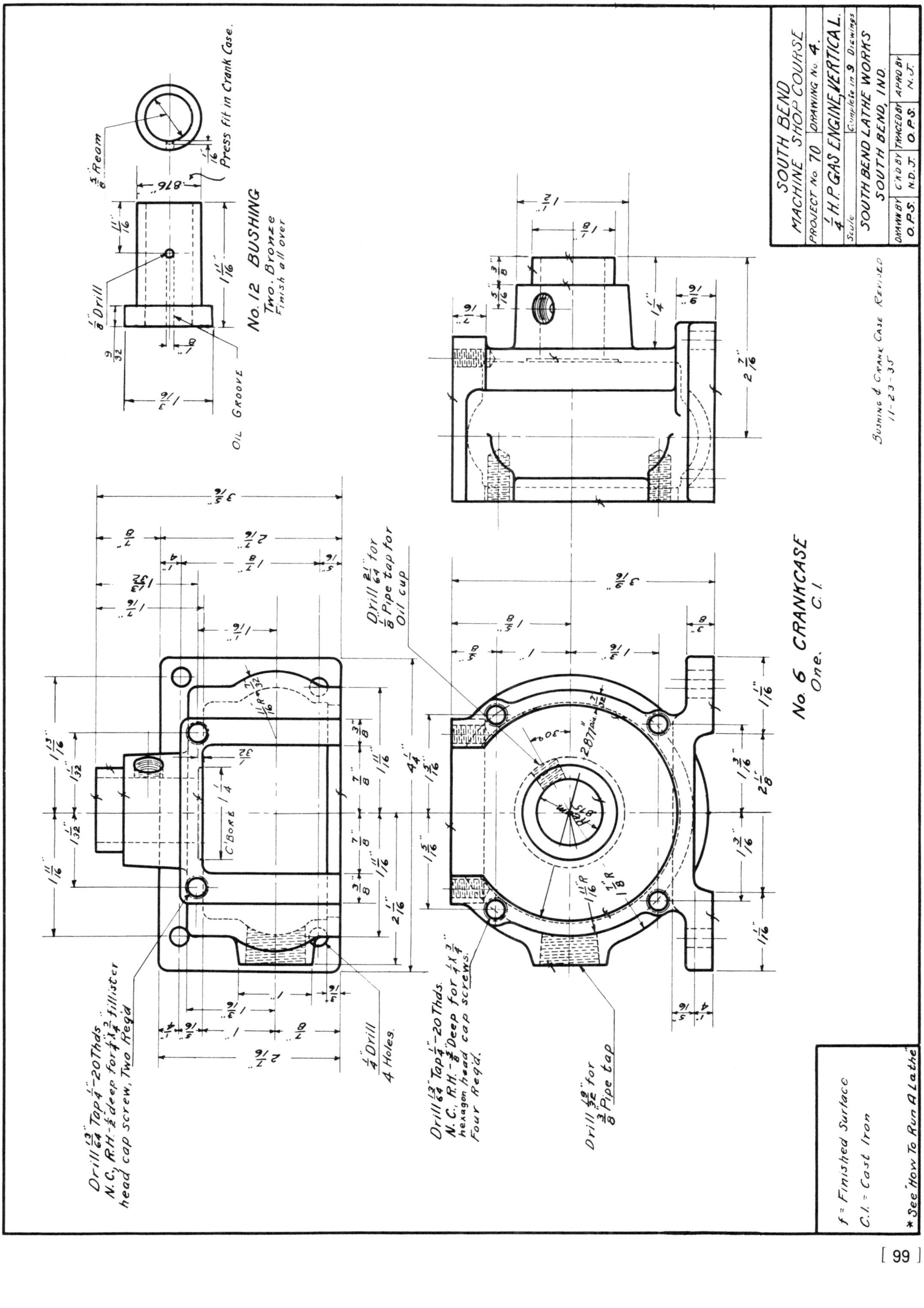

SOUTH BEND MACHINE SHOP COURSE
PROJECT No. 70
DRAWING No. 4.
1/4 H.P. GAS ENGINE, VERTICAL.
Complete in 9 Drawings
Scale
SOUTH BEND LATHE WORKS
SOUTH BEND, IND.
DRAWN BY O.P.S.
C'K'D BY N.D.J.
TRACED BY O.P.S.
APP'D BY N.J.
No. 12 BUSHING
Two. Bronze
Finish all over
5/8" Ream
Press fit in Crank Case.
1/8" Drill
Oil Groove
No. 6 CRANKCASE
One. C.I.
Drill 21/64" for 1/8" Pipe tap for Oil cup
C'Bore
Drill 13/64" Tap 1/4"-20 Thds. N.C. R.H.-1/2" deep for 1/4 x 1/2" fillister head cap screw, Two Reqd.
1/4" Drill 4 Holes.
Drill 13/64" Tap 1/4"-20 Thds. N.C. R.H.-3/8" Deep for 1/4 x 3/4" hexagon head cap screws. Four Reqd.
Drill 19/32" for 3/8" Pipe tap
Bushing & Crank Case Revised 11-23-35
f = Finished Surface
C.I. = Cast Iron
* See "How To Run A Lathe"

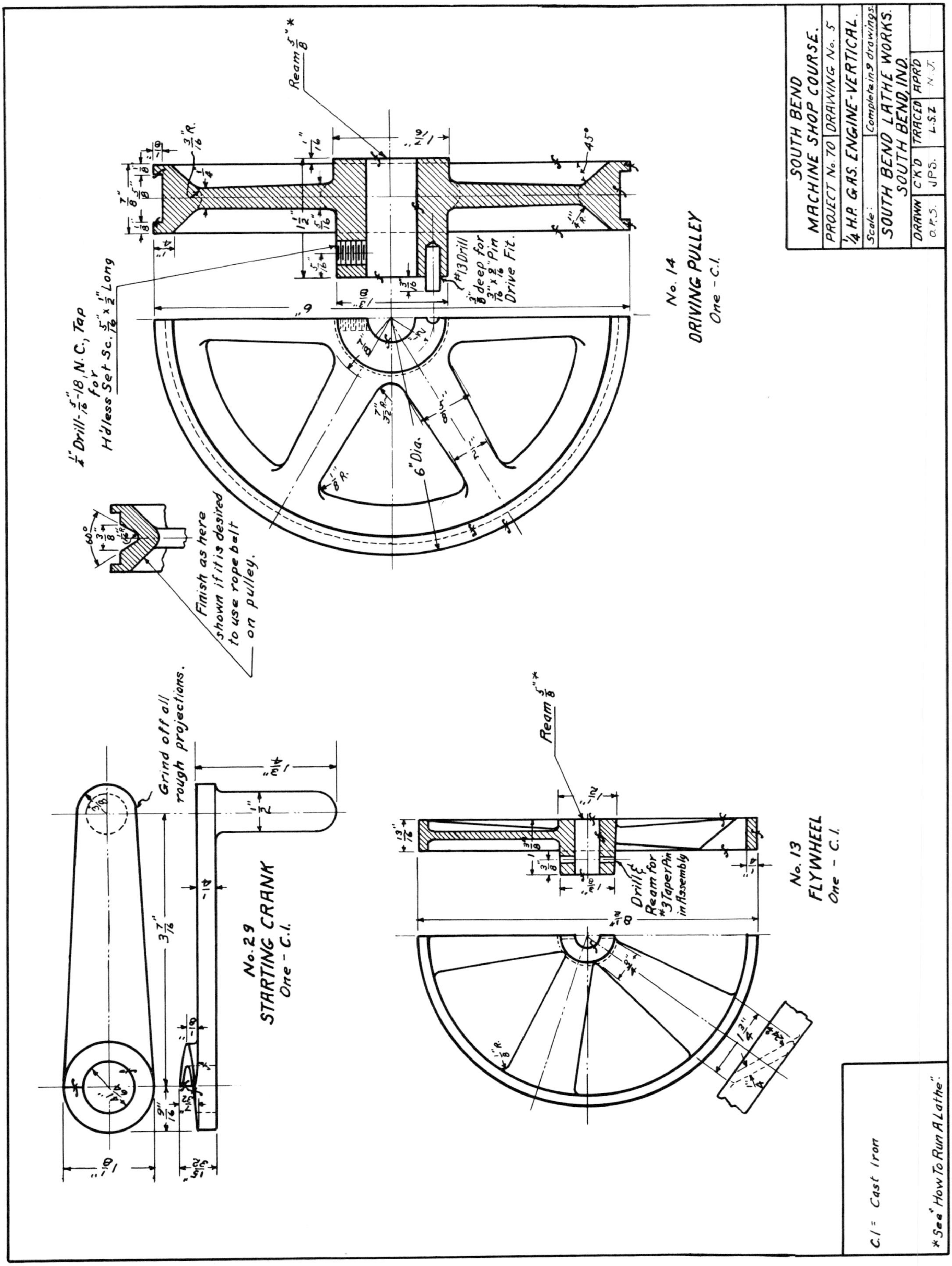
SOUTH BEND MACHINE SHOP COURSE.
PROJECT No. 70
DRAWING No. 5
1/4 H.P. GAS ENGINE-VERTICAL.
Scale:
Completeing drawings.
SOUTH BEND LATHE WORKS. SOUTH BEND, IND.
DRAWN O.P.S.
C'K'D JPS.
TRACED L.S.Z.
APP'D N.J.
Ream 5/8"*
1/4" Drill-5/16"-18 N.C. Tap for H'dless Set Sc. 5/16" x 1/2" Long
#13 Drill 3/8" deep for 3/16" x 2" Pin Drive Fit.
6" Dia.
No. 14
DRIVING PULLEY
One - C.I.
Finish as here shown if it is desired to use rope belt on pulley.
Grind off all rough projections.
No. 29
STARTING CRANK
One - C.I.
Ream 5/8"*
Drill & Ream for #3 Taper Pin in Assembly
No. 13
FLYWHEEL
One - C.I.
C.I = Cast Iron
* See "How To Run A Lathe".

¼ H.P. VERTICAL AIR-COOLED GASOLINE ENGINE (Continued)

OPERATIONS:

JOB No. 17—PART No. 8—CRANK PIN

1. Chuck stock.
2. Face ends to length.
3. Rough turn diameter.
4. Finish turn diameter.
5. File and polish.
6. Cut off, allowing finish on end.
7. Chuck work.
8. Face unfinished end to length.

JOB No. 18—PART Nos. 9 and 10—CRANKSHAFT

NOTE: Make shaft in one piece, the center section between disks to be cut out after assembly.

1. Lay off and center ends of stock.
2. Face ends to length.
3. Rough turn diameters.
4. Finish turn diameters.
5. File and polish.

NOTE: The taper pin hole and the tapped hole for the contact point are to be drilled in assembly.

JOB No. 19—PART No. 30—COMMUTATOR

1. Chuck stock.
2. Face.
3. Drill, bore, and ream.
4. Press a mandrel in work.
5. Face unfinished side to width.
6. Turn diameter.
7. Drill and counterbore.

JOB No. 20—PART No. 25—SPARK BRUSH

1. Lay off and mark outline.
2. Cut around outline with a cold chisel or tin snips.
3. Finish to shape with a file.
4. Drill.
5. Bend curved end.

JOB No. 21—PART No. 28—CONTACT POINT

1. Chuck stock.
2. Turn.
3. Face end and shoulder to length.
4. Cut thread.
5. Cut off work, allowing finish on head.
6. Saw screw driver slot.

NOTE: End of head to be turned on radius with circumference of commutator in assembly.

JOB No. 22—PART No. 26—SPARK BRUSH INSULATOR

1. Plane or file ends to length.
2. Plane or file sides to width.
3. Drill.

JOB No. 23—PART No. 27—SPARK BRUSH SCREW INSULATOR

1. Chuck stock.
2. Face.
3. Center stock for drilling.
4. Drill.
5. Turn diameters.
6. Cut off to length.

JOB No. 24—PART No. 24—SPARK LEVER

1. Chuck casting.
2. Bore and ream.
3. Press a mandrel in casting.
4. Face sides to width.
5. Saw.
6. Drill counterspot and tap.

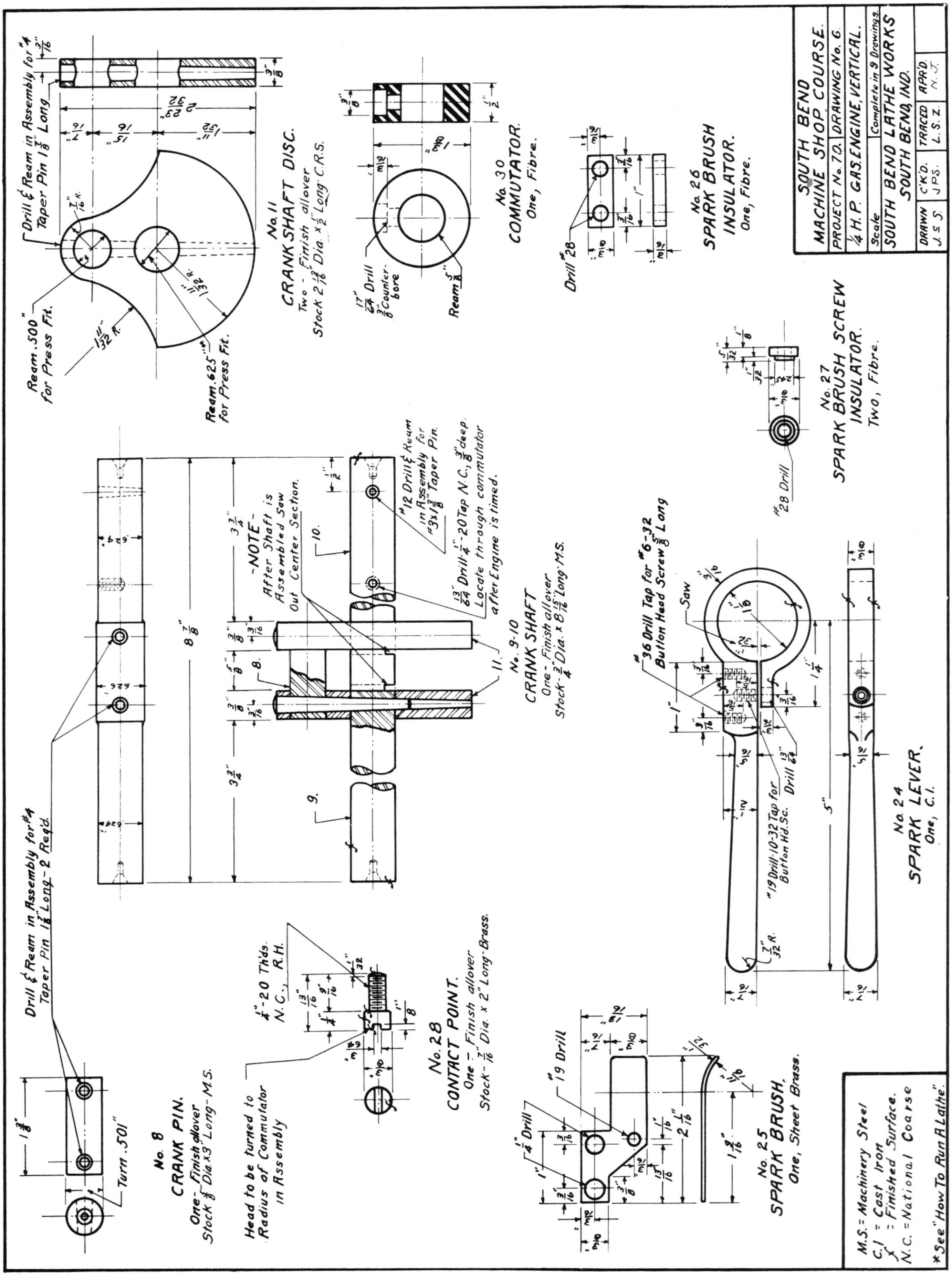
SOUTH BEND
MACHINE SHOP COURSE.
PROJECT No. 70. DRAWING No. 6.
1/4 H.P. GAS ENGINE, VERTICAL.
Scale
Complete in 9 Drawings.
SOUTH BEND LATHE WORKS
SOUTH BEND, IND.
DRAWN J.S.S.
C'K'D. J.P.S.
TRACED L.S.Z.
APP'D. N.J.
No. 11
CRANK SHAFT DISC.
Two - Finish allover
Stock 2 13/16" Dia. x 1/2" Long C.R.S.
Drill & Ream in Assembly for #4 Taper Pin 1 1/8" Long
Ream .500" for Press Fit.
Ream .625" for Press Fit.
No. 30
COMMUTATOR.
One, Fibre.
Ream 5/8"
No. 26
SPARK BRUSH
INSULATOR.
One, Fibre.
Drill #28
No. 27
SPARK BRUSH SCREW
INSULATOR.
Two, Fibre.
#28 Drill
No. 9-10
CRANK SHAFT
One - Finish allover
Stock - 3/4" Dia. x 8 15/16" Long M.S.
-NOTE-
After Shaft is Assembled Saw Out Center Section.
#12 Drill & Ream in Assembly for #3 x 1 3/8" Taper Pin.
13/64" Drill 1/4"-20 Tap N.C., 3/8" deep. Locate through commutator after Engine is timed.
#36 Drill Tap for #6-32 Button Head Screw 5/8" Long
Saw
#19 Drill 10-32 Tap for Button Hd. Sc.
Drill 13/64"
No. 24
SPARK LEVER.
One, C.I.
Drill & Ream in Assembly for #4 Taper Pin 1 1/8" Long - 2 Req'd.
Turn .501"
No. 8
CRANK PIN.
One - Finish allover
Stock 5/8" Dia. x 3" Long M.S.
Head to be turned to Radius of Commutator in Assembly
1/4"-20 Th'ds. N.C., R.H.
No. 28
CONTACT POINT.
One - Finish allover
Stock - 7/16" Dia. x 2" Long - Brass.
#19 Drill
1/4" Drill
No. 25
SPARK BRUSH.
One, Sheet Brass.
M.S. = Machinery Steel
C.I. = Cast Iron
f = Finished Surface.
N.C. = National Coarse
* See "How To Run A Lathe."

¼ H.P. VERTICAL AIR-COOLED GASOLINE ENGINE (Continued)

OPERATIONS:

JOB No. 25—PART No. 15—MIXING VALVE

1. Chuck casting to drill and tap for Gas. Adj. Screw.
2. Face end of boss.
3. Turn diameter of boss.
4. Center casting for drilling.
5. Drill and tap.
6. Cut thread on diameter of boss, fitting to ⅜″—24 thread National Fine Standard Nut.
7. Chuck casting to bore bowl.
8. Face.
9. Bore bowl.
10. Bore air intake.
11. Bore 45° angle valve seat.
12. Drill and tap for ⅛″ and ¼″ pipe.
13. Drill No. 52 between ⅛″ pipe inlet and Gas. Adj. Screw inlet.

JOB No. 26—PART No. 16—MIXING VALVE CAP

1. Chuck casting, gripping on small diameter.
2. Face.
3. Turn diameter.
4. Face shoulder.
5. Center casting for drilling.
6. Drill and tap.
7. Turn casting end for end in chuck.
8. Face end of small diameter.
9. Turn small diameter.
10. Cut thread, fitting to ⅜″—24 thread National Fine Standard Nut.

JOB No. 27—PART No. 17—AIR VALVE

1. Chuck casting, gripping on 19/64″ diameter section.
2. Turn 15/64″ and 43/64″ diameters.
3. Face end and shoulder.
4. Turn 45° angle of valve.
5. Turn work end for end in chuck.
6. Turn 19/64″ diameter.
7. Face end and shoulder.
8. Center casting for drilling.
9. Drill.

JOB No. 28—PART No. 19—AIR ADJUSTING SCREW

1. Chuck stock.
2. Face end.
3. Center drill and support end of work with tailstock center.
4. Turn diameters.
5. Knurl.
6. Cut thread.
7. Turn .125″ diameter section.
8. Cut off work.
9. Face head section to length.

JOB No. 29—PART No. 21—PACKING NUT

1. Chuck stock.
2. Face.
3. Center stock for drilling.
4. Drill, bore, and tap.

JOB No. 30—PART No. 20—GAS ADJUSTING SCREW

1. Chuck stock.
2. Turn diameters.
3. Knurl.
4. Cut thread.
5. Turn ⅛″ diameter section.
6. Turn angle of point.
7. Cut off work.
8. Face head section to length.

JOB No. 31—PART No. 18—SPRING

1. Wind spring in lathe.

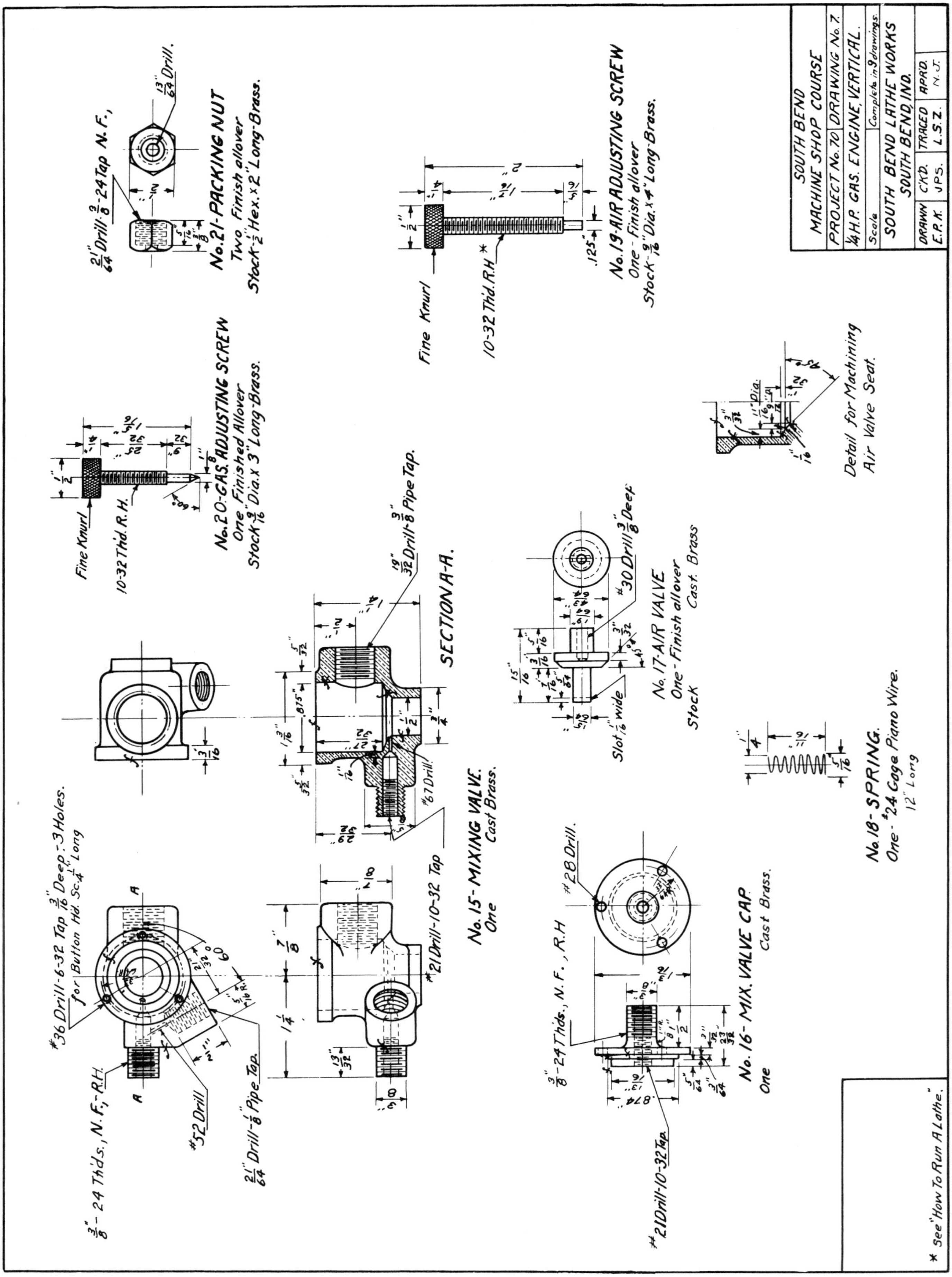

PRINTED IN U. S. A.

C5424—IRXM—11-54